AFRICAN INDEPENDENT CHURCHES TODAY

Kaleidoscope of Afro-Christianity

AFRICAN INDEPENDENT CHURCHES TODAY

Kaleidoscope of Afro-Christianity

Edited by

M C Kitshoff

African Studies
Volume 44

The Edwin Mellen Press
Lewiston/Queenston/Lampeter

Library of Congress Cataloging-in-Publication Data

African independent churches today : kaleidoscope of Afro-Christianity / edited by M.C. Kitshoff.
p. cm. -- (African studies ; v. 44)
Includes bibliographical references and index.
ISBN 0-7734-8782-4 (hc)
1. Independent churches--Africa. 2. Africa--Church history. 3. Africa--Religious life and customs. I. Kitshoff, M. C. (Michiel Casparus) II. Series: African studies (Lewiston, N.Y.) ; v. 44.
BR1360.A534 1996
289.9--dc20
96-34097
CIP

This is volume 44 in the continuing series
African Studies
Volume 44 ISBN 0-7734-8782-4
AS Series ISBN 0-88946-175-9

A CIP catalog record for this book is available from the British Library.

The Edwin Mellen Press
Box 450
Lewiston, New York
USA 14092-0450

The Edwin Mellen Press
Box 67
Queenston, Ontario
CANADA L0S 1L0

The Edwin Mellen Press, Ltd.
Lampeter, Dyfed, Wales
UNITED KINGDOM SA48 7DY

Printed in the United States of America

CONTENTS

Contributors To This Volume

Hans-Jürgen Becken of Stuttgart, Germany, previously Rector of the Lutheran Theological College at Mapumulo, South Africa and Secretary for Africa of the Association of Churches and Missions in South Western Germany at Stuttgart. He has been a co-worker of NERMIC for many years.

Don Boschman is a trained missionary engaged in leadership training, church relations and historical research with AICs in Botswana.

Dumisani Dube is the Director of the Institute of Education and Human Development at the University of Zululand, South Africa.

Steve Edwards is Professor and Head of the Department of Psychology at the University of Zululand, South Africa.

Mike C Kitshoff is Professor and Head of the Department of Systematic Theology and History of Christianity and Dean of the Faculty of Theology, University of Zululand, South Africa.

A W Z Kuzwayo is Deputy University Librarian of the University of Zululand, South Africa.

Bobby Loubser is Professor and Head of the Department of Bibliological Studies, Faculty of Theology, University of Zululand, South Africa.

Bongani Mazibuko was Professor and Head of the Department of Missiology, Science of Religion and Practical Theology of the University of

Zululand, South Africa. He is now occupying the chair of Missiology at the University of Durban-Westville, South Africa.

Skhumbuzo Mfusi, an ex-student of the University of Zululand, is teaching Psychology at the University of Transkei, Umtata.

Hlengiwe Mkhize, previously Senior Lecturer at the University of Zululand, is Senior Lecturer in Psychology at the University of Witwatersrand, Johannesburg, South Africa.

Matthias Mohr, studied at the Philipps-Universität in Marburg, Germany, and holds a doctor's degree from the University of Natal, South Africa.

Hennie Pretorius, previously Rector of the Decoligny Theological Seminary, is now involved in a Bible Training Programme for AICs in the Cape Peninsula, South Africa.

Martin Pauw, who previously taught in Malawi, is Professor in the Department of Missiology, University of Stellenbosch, South Africa.

David A Shank was the Director of Documentation and Resource Center, Group for Religious and Biblical Studies, Abidjan, Côte d'Ivoire. Presently he is a consultant for the Mennonite Board of Missions and Congregational Overseer in Indiana-Michigan Mennonite Conference, United States of America.

Roelie van der Spuy, who hols a doctor's degree, is minister of the Uniting Reformed Church in Africa, Senekal, South Africa.

Attie van Niekerk, Rector of the Stofberg Theological School, Turfloop, also teaches Practical Theology at the University of the North, South Africa.

Danie van Zyl is Co-ordinator of a Bible Training Programme for AICs in the Cape Peninsula, South Africa.

Musa Xulu is Professor and Head of the Department of Music at the University of Zululand, South Africa.

Series Editor's Preface

This volume serves as a tribute to the person and work of Professor G.C. Oosthuizen, leading South African academic and researcher in the field of religion. During the past ten years he was also active as Director of NERMIC, the research unit for New Religious Movements and Independent/Indigenous Churches, attached to the University of Zululand.

NERMIC has collected a sizeable amount of primary material on the African Independent/Indigenous Churches, has encouraged and sponsored scores of researchers to engage in research projects on the AICs, and has published various collections of studies and essays on the AICs. Further studies and oral histories of AICs are awaiting publication. Apart from these volumes edited or co-edited by Professor Oosthuizen, he has also widely written and published on the AICs.

This volume contains studies and essays written by people who know Professor Oosthuizen and who appreciate his work of quality. They are or were co-workers of NERMIC and/or academics of the University of Zululand. The work was initiated, collected and edited by Professor M.C. Kitshoff, Dean of the Faculty of Theology, University of Zululand, and chairman of the Project Committee of NERMIC.

With this volume the contributors want to say that they respect Professor Oosthuizen and acknowledge his sterling work over many years.

Irving Hexham,
Calgary

INTRODUCTION

Afro-Christianity as expressed in the African Independent/Indigenous Churches (AIC) displays a kaleidoscopic picture. This analogy can be refined when we allow the dictionary to remind us that a kaleidoscope is an instrument or toy containing "small brightly-coloured fragments of glass and mirrors which reflect these to form changing patterns". Applying this description to the AIC, key words are "small", "brightly-coloured", "fragments", "reflect" and "changing patterns".

The African Independent/Indigenous Churches are virtually daily increasing in numbers. Today probably 33% or 10 million of the Black population of South Africa see themselves as part of African Christianity as manifested in the AIC. These large numbers are not always evident, though one occasionally gets some idea of this vast movement, especially during annual religious rallies or at funerals of prominent leaders. However, the real life, worship and work of the AIC find expression in *small* groups which meet regularly, groups often consisting of not more than between 20 and 30. It is as if they know that in a smaller group a warmer fellowship, a deeper trust and a stronger release of power are experienced. And such small groups do not need large or costly buildings; they function as "house churches".

And colourful they are - at least in two respects. The members of the AIC often prefer to be clothed in their church uniforms. The women are conspicuous in theirs with white, green and blue as dominant colours.

But the term *brightly-coloured* can also be used metaphorically. The AIC in their customs, orality, beliefs, worship, activities and community involvement, not to mention more, reveal a colourful display of variety, which often leaves the mainline churches pale and dull. This variegation is obvious when key words in recent research topics on the AIC are observed. Here are a few: Traditional religion, magic, transformation, traditional healing, faith healing, symbolism, enculturation, the "Spirit", mutual aid societies, beliefs, practices, music and dance, modernisation process, development, sacraments, ancestors, politics, church names, theological training, leadership, prophets and priests, oral tradition, visions and dreams, revelation, messianism, Scripture use, world outlook, burial rites, sources for writing AIC history, the use of water, hymns, etc.

Apart from the colourful aspects displayed by research on these and many other topics, research on the histories of the many thousands of the AIC will add further colour to the variegated patterns of these churches.

But our knowledge at this stage only has reference to *fragments* of this mighty movement, the AIC. Whereas they display a predominantly holistic view, research by its very nature is fragmental. Moreover, the research field opened up by the AIC is so vast that what has been achieved - and the research achievements and output are impressive - can only be called fragments of the whole.

The AIC *reflect* like the mirrors in the kaleidoscope. They reflect the soul of people seeking ecclesiastical independence, people desiring to express their religious faith in an Africa-oriented manner, people seeking for holistic health, people desiring to be moved by the spirit as they understand and experience it, people who by self-help and mutual aid try to free themselves from their material misery, people who often serve as stabilizing factors in a state of flux and upheaval, people who had sought a place to feel and home and have found it in the AIC.

In every religious movement there is something ancient, something eternal, something above change. The AIC are no exception. But *changing patterns* do reveal themselves in every religious movement, also in the AIC. Modernization and the plastic era have also invaded the AIC. Plastic containers for the powerful healing water, plastic bowls for the offering, and novilon and hide as co-materials for the regalia of the leader can serve as examples of some form of change. Also Western musical instruments have found their way to the AIC.

But these are not incisive changes. The real changes are effected unintentionally in those areas where both African traditionalism and Western-oriented Biblical Christianity meet Afro-Christianity. Millions of people from the African traditional religion have joined the AIC. Influencing and change could not be prevented. On the other hand Biblical Christianity of European-American stamp has not left the AIC untouched. Bible study programmes and theological training activities directed at the AIC contain strong potential for change.

Urbanization, deprivation and unemployment forced them to adapt or die. New schemes were devised for caring and curing, for helping and healing. It remains to be seen how these churches are going to react to further political change and even greater social challenges.

Whatever the different perceptions, views, expectations and opinions of the AIC, nobody can dispute its kaleidoscopic character. But then this volume consisting of studies on various aspects of the AIC cannot be other than also a kaleidoscope in miniature. This volume also contains small brightly-coloured fragments reflecting some changing patterns of the AIC.

But this volume is not only a kaleidoscope, it is also a tribute, a tribute to Professor G C Oosthuizen, for ten years the Director of the research unit for New Religious Movements and Independent Churches (NERMIC), previously sponsored by the Human Science Research Council, and housed at the University of Zululand. Six of the contributors are from the University of Zululand. They honour a colleague worthy of appreciation. A number of the contributors have been engaged in research projects which were partly financially assisted by NERMIC. Their articles for this volume are clear tokens of honour paid by them to Professor Oosthuizen. All the contributors to this volume join in paying homage to an outstanding academic and an untiring researcher. They salute him.

M. C. Kitshoff

PART ONE

ON THE HISTORY OF AFRO-CHRISTIANITY

CHAPTER 1

CONSIDERATIONS ON SOURCES FOR WRITING AFRICAN INDIGENOUS CHURCH HISTORY

by

H. L. Pretorius

African indigenous church (AIC) historiography is not fundamentally different from the writing of the history of the African church at large or from that of any other church. In all cases the "main purpose of the historian is to make a critical and comparative use of his sources in order to produce an intelligent and meaningful description of the past" (Verstraelen-Gilhuis 1982:18).

Apart from all academic considerations, and these are weighty enough, the realization of John R Mott's vision (cf. Irvine 1976:189-90) necessitates the preservation of AIC historical sources: "It is a Christian, that is, a missionary *obligation* to collect, preserve and protect the available data (on missions and the expansion of the Church) so that when the 'younger churches' begin to be interested in their own pre-history this documentation is available". There are ample signs that this interest is growing among many South African indigenous churches.

As far as sources are concerned, AIC historiography should move with developments in the fields of church and secular history. Many valuable lessons and techniques could be learnt from other disciplines.

Given the disadvantage of AIC history over against that of mainline churches, namely, that written sources, let alone archives, are comparatively scarce, the basic problem confronting the researcher is how to disclose a mass of unrecorded AIC history. The first step should be to draw upon those sources which are extant and available. This should be followed by a search for untapped sources.

A concomitant point is to concentrate not on formal institutions and movements which have kept written records but to attend to less formal and often marginalized groups such as grassroots eye-witnesses, resistance and protest groups on local levels, womens' groups, peasants, agricultural and industrial labourers and mineworkers. It should be taken into account that the overwhelming majority of AIC members in South Africa belong to the latter sections of society.

In other words, different versions, based on different types of sources are necessary to arrive at a more satisfactory and a firmer historical basis of the events that took place.

It could even tentatively be inferred that when writing church history, written evidence will normally be supplemented by oral sources. The opposite would be the case when writing AIC history: oral sources will usually be more important than literary ones.

Whether written or verbal sources or both are used, sources in the vernacular have a vital role to play and should be used with circumspection in addition to those in Western languages.

Literary sources

As for literary sources, figures available lead to the generalized conclusion that published materials on AICs are relatively scarce in comparison with those of mainline churches and missionary societies. This is appreciable when taking into consideration that the majority of the membership and many leaders, especially in the formative years, were not necessarily literate or were semiliterate.

The result was that not much written or published material was produced by leaders and members of AICs themselves. And for material that was produced archives or other preservation facilities were not readily available. These remarks are made in spite of some notable exceptions of leaders who did have the insight and persistence to gather and preserve sources and to use them to produce histories of their churches.

The search for and identification of further literary sources extant among AICs themselves should be continued vigorously because valuable historical evidence could be brought to light in this manner.

Together with other categories of unpublished and published materials generally available in archives, libraries, law reports, other collections and novels, the harvest of written sources might prove to be not as scant as is often supposed. Kamphausen, among others, has shown that the possibility of discovering new, undisclosed literary sources in archives and elsewhere does exist. A systematic probe for literary sources in overseas institutions such as government, church and missionary archives particularly in the United Kingdom, Germany and the United States of America needs to be undertaken.

Furthermore, an expanded bibliography specifically on the history of South African indigenous churches could be a most useful historiographical tool. The compilation and continuing up-dating of such a bibliography seems to be an urgent starting point and a task which is long overdue.

Apart from the identification of written sources on AICs, their collection and preservation is a further problem which has to be overcome. It has been pointed out that the collection, housing and documentation, in fact, the very consolidation of archival sources, constitutes a major challenge to African historiography.

How is one to assess the present situation regarding the preservation of literary sources for the writing of South African indigenous church history?

The idea of one centralized collection of all or as many as possible literary sources on these churches (akin to the Harold Turner Collection at the Interact Research Centre - formerly Centre for New Religious Movements - Selly Oak Colleges, Birmingham) is hardly feasible although not impossible.

Such a collection need not necessarily contain original sources. Modern duplicating facilities could provide at least copies of the majority of literary sources. Sources related to history should be readily identifiable and accessible within the total collection.

A second option would be a central data base which contains the necessary information on the location of specific sources. Such a data base would have to be accurate, detailed and regularly updated.

The third possibility is that individual researchers would have to initiate their own inquiries by starting at the institute at which the required source is most likely to be found.

It is suggested that documentation concerning AIC historical sources in South Africa is not dealt with on a co-ordinated basis. Such a situation does not satisfactorily promote research, and ways and means need to be sought to improve the position. The establishment of a task group or a consultative body to co-ordinate information on sources would be ideal but such a suggestion would most likely not be realistic in the present academic and financial climate.

A more viable possibility is the setting up of a co-ordinating agency on a permanent basis at an existing institution. The main aim of such an agency should be to co-ordinate continually updated information where specific documentation on the history of indigenous churches in South Africa is available. To fulfil its co-ordinating function an agency of this kind would have to be acceptable and useful to all ecclesiastical traditions, academic institutions and, above all, the AICs themselves.

Oral Sources

Apart from written sources, oral sources cannot be ignored by AIC historiographers. As early as the mid-1960s this has been an emerging theme in African history. Taking into account the absence of archives and the paucity of literary sources for South African indigenous church history, one can assume that oral sources have much to contribute.

It is indeed fortunate that black African cultures have their own indigenous form of storing and passing on of history: oral tradition. In fact, only in modern times have written versions exceeded oral sources as a method of reconstructing the past. Oral transmissions were even responsible for the normative texts of the Christian tradition. The use of such sources in producing AIC histories would thus be in agreement with ancient Christian usage.

Since the 1970s much has been done in South Africa for the systematic collection of oral testimony. A few significant examples of Southern African studies reflect the fruit of this approach applied to the history of indigenous churches.

For South African case studies reference to oral history is usually more appropriate than oral tradition, as the former is mainly used with regard to contemporary events and situations and the latter reaches further back, at least beyond a generation. On the same basis one would also have to distinguish between oral sources as well.

Owing largely to the comparative scarcity and often one-sidedness of documentary source materials, oral transmissions have become generally accepted as a legitimate source for the study of the past. Oral historiography is, however, a complex practice. The solving of a number of problems in this field remains a formidable task, but a task worthwhile to open up unique possibilities which the use of other historical sources does not offer.

Many of the difficulties of using oral sources are typical of all history writing. Whatever the type of source used, problems relating to its availability, reliability and interpretation have to be dealt with. In addition, oral sources present particular challenges in the area of interviews and popularization. These five categories will now consecutively be considered and applied to South African indigenous churches.

As for the *availability* of sources the major problem is that significant historical data are irreplaceable when the bearers of this information impart only part of it by word of mouth to the next generation without any written sources being available. Many indigenous church leaders have passed away without having transmitted satisfactory oral histories of their churches of which they were often the founders. It is here that the historical researcher could be of great benefit to the AICs themselves as well as to the academic world, rescuing these oral sources from extinction by recording them.

In South Africa, as elsewhere, the collection of oral evidence could be hampered both by modern bureaucracy and by traditional suspicion towards the questioning outsiders. The second problem is alleviated by the general keenness of the AIC to have their histories recorded, and by a relationship of trust towards the person who does the research.

Experts in the field of oral sources have emphasized that not only the interpretations of field researchers should be made available but also their recording and the necessary data for their evaluation. Although not always feasible, such a practice has an important advantage: Historiography develops not only through the disclosing of new evidence, but also through the reconsideration of older evidence. All oral evidence available should be part of this process.

With reference to *reliability* it should be noted that all sources, including AIC oral sources, could be of unsound and inconsistent quality. As this reduces their value, they should as far as possible be tested through comparison with independent sources to discover the degree of their reliability. Proof of reliability is given only when at least two independent sources corroborate the evidence of the same event or situation. If only one witness is available it is not considered to be reliable in that the events or situation could have been invented or misrepresented beyond recognition.

Oral sources used in conjunction with written documents can be most profitable. Writings and verbal sources could offer mutual confirmation of each other without either of them being inferior. When they converge, both are part of the proof and a higher degree of reliability has been achieved than if only one of the two was available. The following procedure should prove to be valuable:

> After the collection of both written and oral source, the data have to be checked again and again: from the written sources to the fieldwork data and back again. Wherever possible the archival - and in most cases but not exclusively European - sources have to be supplemented if not corrected by oral sources, whereas the field work data have to be questioned from the documents (Verstraelen-Gilhuis 1982:20).

Various forms of distortion can take place when the researcher relies on oral sources. An example of how reliability of AIC sources can be affected is the lack of distinction between mythical and historical data. Myths should be viewed metaphorically as summaries or interpretations of past events, and should not be taken literally as if they are narrations of events that took place.

A high degree of reliability in reconstructing history can be achieved through the use of oral sources. To attain this, independent sources, both verbal or literary, will have to be utilized. The scarcity of written sources to cross-check the verbal evidence will remain a difficult problem for AIC historiography.

Interviews aim at the recording of oral history. Questionnaires which usually require less effort and time to administer are useful, but do not equal the value of recorded conversations. The latter facilitate personal interaction which has a number of benefits. The development of a sound relationship of mutual trust between the interviewer and interviewee as well as the creation of a sense of partnership is highly rewarding in terms of developing a situation in which frank exchange takes place and accurate information is maximally imparted. Complex and subtle problems can be addressed while distortions are being corrected. Time should be spent explaining the purpose of the research and winning the confidence of the interviewee and the co-operation of the church concerned.

The extraordinary value of "participant observation" in the case of AICs has been proved empirically by researchers such as Turner and Daneel. This does not apply only to research of a comprehensive nature but, among others, also to the tapping of oral sources by way of conversing with people about the histories of their churches. Obviously knowledge of the society and the circumstances of the interviewee or his/her church and community is necessary to know which questions to ask and which not to ask.

When, as in the multilingual South African situation, interviews are conducted by means of an interpreter, the interviewer will have to be especially careful of distortions. To minimize the loss of crucial particulars or nuances, a detailed rather than an impressionistic translation from a vernacular language should be used. The ideal would obviously be for the researcher to be fluent in the language of the interviewee.

The conducting of follow-up interviews in which a more open and relaxed situation prevails will enhance the effectivity of the researcher's efforts.

The *interpretation* of historical data is no easy task, irrespective of whether literary or oral sources or both are used. Some exceptional difficulties which oral sources create for the interpretation of historical information need to be taken seriously by the AIC researcher. Vansina (1988:191,194-6) has not only described the nature of these difficulties but has also suggested ways of managing them.

When the interpretation of data about the past is done, certain information which is not considered to be significant for the present is discarded. The remainder of the information which is selected as having meaning is then through interpretations, altered to give it a new meaning. Both the processes of selection and interpretation are tied to messages. These messages are characterized by a double subjectivity: that of the sender of the message in whose mind contemporary events and situations are interpreted (encoded); and that of the receiver (who could be the historian) in whose mind another interpretation of the message takes place (decoding). Subjectivity is reduced when sources are tangible, as written sources are. Usually no alteration of the sources takes place. On the other hand, in the case of intangible sources such as oral ones, interpretation accumulates as the sources have to be reproduced from time to time until they are recorded.

The *popularization* of history has a direct bearing on oral sources. Although a critical stance is always appropriate, the involvement of ordinary people with interest in and appreciation for history should not be seen as a threat or disadvantage. On the contrary, "history from below" implies both an alternative and a challenge to official views of history and could promote significantly amended and more comprehensive interpretations.

This possibility is important when considering South African indigenous churches in whose case oral sources, the voices of ordinary members and leaders, are often providing the major part of the data necessary to write their histories. The very process of collecting oral evidence from them encourages a consciousness of and a pride in their own history. Thus the makers of AIC history are involved in the writing of this history even if it be through the historian. In other words, the researcher will assist the AICs discover their own history. The benefits of such a development for church historiography could be immense. It is important that the results of the research be made available to the church concerned even if all members are not able to read the language of the publication.

In conclusion: as a result of the scarcity of written sources and the presumed massive availability of oral sources, the latter will have to be used more widely in indigenous church history. The assessment made some years ago that "in South Africa very few projects have attempted to use oral historical data" (Saunders & Derricourt 1974:xi) is hopefully being redressed.

Despite the limitations and risks oral sources are indispensable for writing AIC history. Wherever possible, available written sources on indigenous churches should supplement oral ones, for each type has its own particular value for historiography. This pattern, rather than the other way round, will most likely be prominent in the majority of the histories of South African indigenous churches.

The consideration and reconsideration of known data on AICs as well as the search for new data from known and previously unknown sources is, therefore, part of an ongoing process. Within the broader scope of African church history this process has been well developed during the last few years. One can agree with the following statement:

> Most of the new style publications on church and mission history are to a great extent based on newly discovered archival sources and the reinterpretation of the existing ones within a broader historical framework and from an African point of view. In the interest of recovering the African perspective, all possible resources, oral and written, have to be exploited - with all the risks involved, but at the same time making use of all the safeguards at one's disposal (Verstraelen-Gilhuis 1982:21).

The establishment of ongoing structures for the collection of oral sources deserves serious consideration. Though this seems to be a massive task, it would provide an opportunity to preserve significant oral information and prevent it from falling into disregard as has sadly happened to so many written sources.

Iconographical Sources

Iconographical sources are the last type to be considered, though briefly because this is an underdeveloped area.

It is not difficult to establish that AICs, as has been illustrated in a few significant publications, (cf. Oosthuizen 1979) do have a rich iconography. The appreciation and understanding of this conglomerate of "icons" could still be the key to reveal many an enigma in the field of AIC research, including historical dimensions.

It is therefore encouraging to note that the use of photographic visual sources of African history, for instance, is elsewhere enjoying growing interest. This does not seem to be the case in South Africa as far as indigenous churches are concerned. In spite of a few pioneering publications, the debate on this theme scarcely seems to have begun. In an initial stage the assistance of art historians should be sought to document, index and catalogue extant photographic materials.

Although the sources available for writing South African indigenous church history are scarce in comparison with those at the disposal of mainline churches, a wealth of sources - literary, iconographic but especially oral - have as yet not been tapped. The exploitation of untapped sources promises rich gains for South African church historiography.

BIBLIOGRAPHY

Irvine, C 1976. The documentation of mission in the nineteenth and twentieth centuries. *Missiology* 4:2.

Kamphausen, E 1976. *Anfänge der kirchlichen Unabhängigkeitsbewegung in Südafrika*. Frankfurt/M:Lang.

Oosthuizen, G C 1979. *Afro-Christian religions*. Leiden: Brill.

Saunders, C & R Derricourt (eds). 1974. *Beyond the Cape Frontier*. London: Longman.

Vansina, J 1988. *Oral tradition as history*. London: James Currey.

Verstraelen-Gilhuis, G 1982. *From Dutch Mission Church to Reformed Church in Zambia*. Franeker: Wever.

CHAPTER 2

AFRICAN INDEPENDENT CHURCHES IN MALAWI: BACKGROUND AND HISTORICAL DEVELOPMENT

by

C. M. Pauw

The phenomenon of religious independency in Africa, particularly in the form of independent or indigenous African churches, has been the subject of scrutiny and debate for a long time. The first signs of independency already made their appearance before the turn of the century, particularly in the form of what was called Ethiopianism. In South Africa this was taken note of in various ways (cf Pretorius 1975, 234 ff) both in government reports and in the writings of several researchers. However, it was to be the seminal work of Bengt Sundkler, *Bantu Prophets in South Africa* (1948, revised edition 1961) that brought the issue out into the open and elicited a growing stream of studies and comments.

The publications of David Barrett brought home both the extent of this movement and its growing impact (1968(a), 1968(b) & 1971). Whereas in 1968 he could make reference to 6000 such movements (admitting that this was not the full number) his later projections indicated the existence of 12 000 such movements in Africa by 1985 with a combined membership of 33 million (1982, 815). A very large portion of these movements has been identified in South Africa (currently estimated to be between 3500 and 4000) with a combined membership of six to eight million or 32% of the total black African population (cf Kritzinger 1988, 28-29). Nevertheless, in every other African country movements have been identified and a stream of literature describing various aspects and factors affecting the phenomenon continues unabated. Turner's bibliography (1979) has long been overtaken by subsequent writings.

The studies appearing on this indigenous movement approach the matter from a variety of angles and presuppositions. They range from a study of the causative factors (Oosthuizen 1969:209-221; Taryor 1976, 95-104), to attempts to develop a typology (Turner, 1967, 1979; Parratt 1979, Chakanza 1982, Gilliland 1986), or to identify trends and themes in the movement (Kiernan 1981, Daneel 1982).

One thing must be evident to any student of the phenomenon. Although a variety of common underlying causes can be identified, and although there is a possibility of classifying churches according to some kind of typology, this must be done with care. There is certainly a danger of generalising and of forcing groups to fit into a certain pattern to such an extent that the result may even be somewhat caricaturistic. The individual *Zits im Leben* of each group or movement, the particular history of a country, both politically and ecclesiastically, particular circumstances socially, ethnically and in terms of the activities and policies of missions and even individual missionaries or leaders of mainline churches, all play a role. In addition to that, each movement has its own leadership personalities and idiosyncrasies, is affected and influenced by other movements and carries within it a dynamic for continuing change as well as in many cases a dynamic for fission and further separation (cf Daneel 1982).

Thus any attempt to classify and categorise, especially if it is done on a continental or even subcontinental scale, will run the risk of product models which are not generally applicable. Malawi (formerly Nyasaland) is a case in point: "The various typologies of independent religious movements which have been suggested from time to time are not greatly helpful for the classification of such movements in Nyasaland" writes Parratt (1979:185). No great prophet-type of leaders of the stature of a Shembe (Oosthuizen 1976) a Kimbangu

(Martin 1975) a Lekganyane (Hanekom 1975) or even a Lenshina (Taylor and Lehmann 1961) have emerged, except perhaps Kenan Kamwana (McCracken 1977: 184; 189:203-208). Even John Chilembwe, the first independent church leader to gain prominency and to date the most written about, founder of the Providence Industrial Mission (cf Shepperson and Price 1958) does not really fit into this model. Neither has he or any other leaders acquired a kind of messianic status as has been the case in other countries. Moreover, the groups found in Malawi do not fall easily into categories such as "Zionist" or "prophet-leaders". Even the "Ethiopian" movements in Malawi during the first decades of the century do not really fit into the type of Ehtiopianism such as described by Sundkler (1961; cf Parratt 1979:185).

BACKGROUND TO RELIGIOUS INDEPENDENCY IN MALAWI

1. A profile of missions and churches

Although it is quite impossible to draw an adequate picture of this background in a short article such as this, it is necessary to briefly point out a few features which are of significance in understanding and analysing religious independency.

Apart from the abortive attempt of the Universities' Mission to Central Africa in 1861, the first mission to enter Malawi did so in 1875. This was the Free Church of Scotland Livingstonia Mission which first settled at Cape Maclear, later moved up to Bandawe on the western Lake Shore near Chintheche and finally became established at Khondowe from where its influence spread all over the Northern Region (cf McCracken 1977). A year later the Blantyre Mission was set up as an enterprise of the established Church of Scotland and in due course extended its sphere of work over most of what is today the Southern Region (cf Ross 1968). With the present-day Central Region still unoccupied, the Dutch Reformed Church in South Africa responded to an invitation from James Stewart of Lovedale in South African who had also had a close involvement with the Livingstonia Mission, as well as an invitation from the Free Church of Scotland, and started at Mvera in 1889. The mission moved its headquarters to Nkhoma in 1912 (cf. Pauw 1980). Thus Nkhoma Mission came to be the third component of three Reformed Presbyterian Missions out of the work of which the Church of Central Africa Presbyterian was later formed. In 1924 Livingstonia and Blantyre constituted as two presbyteries of the newly formed CCAP and in 1926 Nkhoma joined as a third (Pauw 1980:265-282).

Meanwhile the Anglican Church returned to the country when the UMCA re-entered in 1881 and established centres of work at Likoma island and around the southern and western Lake Shore areas (Anderson - Morshead 1955, cf Elston 1972). The work of the Roman Catholic Church only really got under way by 1902 and it gradually spread over the whole of the country (Pauw 1980:34-37).

In addition to the larger missions, a variety of smaller missions, including the South Africa General Mission, the Seventh Day Adventists and several Industrial Missions were already present in the country by the turn of the century. They were mainly restricted to certain areas in the Southern Region with a concentration in the Mulanje district and to some extent the Ntcheu district. In the history of no less than seven of these missions the name of Joseph Booth is prominent (Chakanza 1987:135-145). Described by Shepperson and Price as a religious hitchhiker his activities created frustrations for most other missions, but nevertheless led to the formation of a variety of churches including a considerable number of independent, African led churches (cf Shepperson and Price 1958:18-36; Wishlade 1965:12ff). Both John Chilembwe who established the Providence Industrial Mission, and Elliot Kamwana Chirwa who introduced the Watch Tower Society into the country in 1908, were protégés of Booth.

A fairly clear picture emerges by the end of the first decade of this century. As far as (Western) missions are concerned, a clear-cut division of areas of labour had been established between the three Reformed Presbyterian Missions. The Anglicans worked mainly around the southern and western parts of the Lake Shore and the Catholics were gradually spreading out over the entire country. Close co-operation between the Zambezi Industrial Missions, the Nyasa Industrial Mission, the SAGM and the three Presbyterian Missions was already established by 1910. At the third General Mission Conference held at Mvera in that year a Consultative Board of Federated Missions was set up which continued to function right up to 1967 and which was instrumental in the establishment of the Nyasaland Christian Council in 1939, later to become the Christian Council of Malawi (Pauw 1980:38-42). These six member bodies of the Consultative Board agreed to a basis of co-operation which included recognition of respective church membership, common standards of religious instruction for the catechumenate and an agreement as to the respective spheres of work of each mission (cf Report of Third General Missionary Conference 1910:79). This helped to avoid friction and unhealthy competition. The fact that most of the larger Protestant bodies could co-operate in this way (Anglicans, Adventists, and several smaller groups, however, did not participate in the Consultative Board) ensured a greater degree of co-ordination and unity and it may be assumed that this was a factor in the lower (or later) incidence of African Independent Churches in certain areas, notably within the spheres of Blantyre and Nkhoma Missions.

As far as missions from outside the country are concerned, the situation as it existed by 1910 remained basically unchanged for half a century. No significant new groups or missions entered the country and the churches developing from these older missions could make good progress. The CCAP became by far the largest Protestant body in the country, with a total of 188 000 communicant members by 1964, the year of national independence.

Independence for the country of Malawi, however, signalled a new development. From the early 1960s onward an influx of new missions into the country began to take place which was to bring about a completely different situation as far as the church scene was concerned. Coming mainly from the United States of America, new missions were set up by several groups, including the Southern Baptists, Assemblies of God, and Lutherans from the Missouri Synod, to mention but a few. More missions followed mainly of Pentecostal or Apostolic bent. During the 1980s this was followed by a growing number of evangelistic or revival groups, operating at random throughout the country (cf Kachaje 1990; Van Dijk 1988).

This development served to accelerate the process of schism and fission within the sphere of the African Independent Church movement. A considerable number of new groups emerged of which several have developed links with outside bodies, receiving funds and visitors from countries such as the USA, Britain, South Africa and elsewhere. One effect of this development has been to strengthen the fissiparous characteristic within the African independent movement.

2. Independent churches in Malawi

2.1 The beginning

Barely twenty-five years after Christianity was finally introduced into Malawi, the first independent church already made its appearance. This was the Ajawa Providence Mission, later better known as the *Providence Industrial Mission* which was started by John Chilembwe, a former servant and close associate of Joseph Booth. In 1897 Chilembwe accompanied Booth to the United States where he received training at the Virginia Theological Seminary run by the Virginia State Baptist Convention. In 1900 he returned to Malawi to set up his own mission and school near Blantyre, developing his work along the lines of other Industrial Missions in the country. The history of the PIM and the resistance against the British Colonial authorities which led to the ill-fated, but significant rising of 1915, has been well documented (cf Chakanza 1990:95-106).

Chilembwe, however, was no messiah or prophet-type Ethiopianist leader, even though the movement did draw some followers of less orthodox independent African Churches, particularly those who had been influenced by the Watch Tower prophecy of the final judgment which would come to pass in 1914 (Shepperson 1968:254f).

The rebellion ended in failure, and resulted in the death of Chilembwe and a large number of his collaborators as well as in the demolition of his impressive church building (Pachai 1973:214-224; Shepperson and Price 1958:311). The PIM was able to revive when Dr Daniel Sharpe Malikebu (1890-1978), who had left the country in 1905 to study medicine and theology in the USA, returned in 1926 as a fully qualified medical missionary. He was permitted to re-open the PIM and became its leader for many years. It operated closely with the National Baptist Convention, a Negro Church in the United States which regarded Malikebu as its missionary. Changes in the church name as well as several secessions and splits took place over the years (cf Chakanza 1983:3; 22; 48; 54; 56-57, Parratt 1990:119-129), but the church continues to exist.

2.2 Advent of the Watch Tower Movement

A second movement, associated with early Ethiopianism in Malawi, arose in the Northern Region in 1908 within the precincts of Livingstonia Synod. Elliot Kamwana Chirwa was trained as an evangelist in the Overtoun Institution at Livingstonia. In 1901 he left the mission and after several contacts with Booth who introduced him to the teachings of the Watch Tower Movement, he returned to Malawi in 1908. Within months he had baptised 9000 converts mainly in the Tongaland area, before being deported by the authorities (McCracken 1977:184; 189:201-208). During the following decades a variety of Watch Tower Movements developed all over Central Africa (Shepperson 1968:255). In the Northern Region of Malawi several secessions subsequently took place, notably the *Last Church of God and his Christ* of Jordan Msumba at Chintheche in 1925 (Parratt 1979:191, Chakanza 1983:51f).

Several other independent churches were in due course formed which in one way or another were connected to Booth's activities (Parratt 1979:187-191). Chakanza (1987:136) has identified at least 35 such churches of which 86% are found in Southern Malawi and the rest in the North.

2.3 A new wave of independency

Between 1924 and 1935 a second thrust of the independent movement can be identified in Malawi. Some of these churches were introduced from outside the country. Well known and highly respected amongst these was the *African Methodist Episcopal Church* (cf Sundkler 1961:40-43 for a brief historical outline) which originated in the United States of America already in 1816. It was introduced into Malawi by Hanock Msokera Phiri who had formerly been a member of the Livingstonia Mission and had become acquainted with the AMEC whilst working in South Africa (Parratt 1979:192). In 1924 he established the AMEC at Kasungu from where it spread to various other areas.

A year earlier had seen the introduction of Zionism into Malawi when the American *Christian Catholic Apostolic Church in Zion* appeared on the scene (Parratt 1979:192-195). Over the years this movement gave rise to a variety of Apostolic and Zionist groups, while other Zionist groups were also introduced into the country by migrants returning from Zimbabwe and especially South Africa. Well known today are the so-called spirit-type churches such as the Apostles of Yohane Maranke and of Yohane Masowe which originated from Zimbabwe (respectively the African Apostolic Church of Yohane Maranke and the Apostolic Sabbath Church of God of Yohane Masowe, cf Daneel 1971:315-349). Historically these, as are a variety of other Zionist and Apostolic Churches, are related to the Zionist movement in South Africa and through it with Zion City Illinois, USA.

The most significant group of independent churches to develop during the late 1920s and early 1930s was the group which seceded from the membership of the Livingstonia Presbytery of the CCAP. The first of these was the *African National Church* founded between 1927 and 1929 by a group of Livingstonia trained men, including such as Jordan Msumba, Robert Sambo, Paddy Nyasulu and Simon K Mkandawire (Chakanza 1983:17-18). Robert Sambo is claimed to have introduced this movement into the Central Region. One of the main branches of this church today in the Central Region is the African Abraham Church under Bishop Zerias Numeri Nkomba with headquarters at Chamchele Mission near Lilongwe (personal interviews, October 1990). In this church children and adults are baptised, polygamy is allowed, but restricted to two wives in the case of office bearers, beer drinking is permitted but traditional initiation rites are forbidden as is participation in traditional Nyau ceremonies and dancing.

In the north three further secessions took place between the years 1932 and 1934 (McCracken 1977:273-276). Y M Mkandawire, an ordained minister of the Livingstonia Presbytery, was deposed in 1932 for taking protective medicine against poisoning. Shortly after that he proceeded to form the *African Reformed Presbyterian Church.* The following year YZ Mwasi, the first African Moderator of Presbytery, withdrew on grounds of what was basically a conflict with the way certain missionaries exercised authority in the church (Parratt 1979:198; McCracken 1977:276f).

He formed the *Blackman's Church which is in Tongaland* (Parratt 1978:193-206; cf Mwasi 1933). In 1934 the third breach took place when the Reverend Charles C Chinula, in protest of what he regarded as a harsh and legalistic application of discipline (in his case concerning a charge of adultery), formed the *Eklesia Lanangwa* - the Church of Freedom.

A year later these three united to form the *Mpingo wa Afipa mu Afrika* or Blackman's Church in Africa. This church has continued to exist to the present, but has neither been successful in gaining membership in the Christian Council of Malawi nor in having candidates accepted for training at the Zomba Theological College of the CCAP. Attempts to have candidates trained for the ministry through the Theological Education by Extension programme in Malawi (TEEM), though welcomed by the latter, have not been very successful (Thompson 1987:162-166). Two years before his death in 1970, Chinula was received back into the CCAP.

Apart from smaller secessions within the existing independent churches, the next three decades did not produce any significant new groups, whether by way of secessions from mainline churches or through the spontaneous formation of new indigenous churches.
What is of significance is that by the time the country became independent, the phenomenon of church independency, though firmly established in the north as well as in the south, was still of relative insignificance in the Central Region where the Nkhoma Synod of the CCAP had its sphere of work. A number of independent churches had spread to the Central Region from other parts of the country, but with the exception of the PIM their support was insignificant. Apart from these there was hardly a single secessionist or independent movement which had

originated within the area itself. I have dealt in more detail with this phenomenon, and with reasons for it, in a paper read at a conference of the unit for the study of New Religious Movements and Independent Churches (NERMIC) held in Johannesburg in July 1991. It is, therefore, not necessary to discuss it in detail here.

Suffice to say that the decades since national independence have brought a significant change in the situation, particularly as far as the Central Region is concerned. A brief research visit undertaken by the author towards the end of 1990 has confirmed this trend. A large number of new groups have been introduced to this area, particularly during the 1980s. These include, amongst others, Zionist and Apostolic churches and various offshoots of the African National Church formed in the north during the 1920s. Of the latter the *African Abraham Church* under Bishop Nkomba has spread out over a large area, with 22 "missions" under the care of a "Reverend", each of whom oversees from four to ten pastors ("abusa") who are responsible for local worshipping communities. It is especially in rural areas where this church has gained growing support.

A new movement arising after national independence has already been referred to. Beginning as revival or fellowship groups or evangelistic teams several of these groups have developed into churches, mostly showing Pentecostal or Charismatic characteristics (Kachaje 1990:1-13). Some reveal a negative attitude towards mainline churches, while others establish links with foreign, often conservative, or even fundamentalistic organisations coming mostly from North America. Young people and even professional and educated people play a significant role in several of these movements. In some cases preaching is even done by children in their pre-teens (Van Dijk 1988). Certain groups reveal a negative tendency towards elements of African culture, make use mainly of Western forms of preaching, singing and music and model their organisations on the pattern of similar movements in the West. While some tend to develop into churches, others, in some cases organised by ministers of mainline churches (eg. the CCAP), function as revival teams within the church.

By the end of the 1980s the independence movement already had a following of over one million people. Where Barrett had estimated that 2,9% of the population belonged to independent churches (1982:469): Chakanza already in 1986 estimated the total membership of at least 200 existing independent churches to be 950 000, which then amounted to 11% of the whole population (Chakanza 1987:134).

It is clear that a situation is now developing whereby mainline churches will need to reconsider their attitude to at least some of these churches. Pressure is mounting as more and more seek ecclesiastical recognition such as by being accepted into the Christian Council of Malawi or by being granted avenues through which to acquire better training for their leaders (cf Thompson 1987). A variety of para-church organisations and in a few cases individuals from mainline churches are beginning to render assistance in various ways (cf Chakanza 1987) and this may pave the way for greater ecclesiastical recognition. As is the case all over Africa, it is becoming clear that in Malawi also the African Independent Churches are there to stay.

BIBLIOGRAPHY

Anderson-Morshead, AEM 1955. *The history of the Universities' Mission to Central Africa* Vol I:1857-1909, 6th ed.

Baëta, C G (ed) 1968. *Christianity in tropical Africa.* London: Oxford.

Barrett, D B 1968(a). *Schism and renewal in Africa.* London: Oxford.

Barrett, D B 1968(b). Church Growth and independency as organic phenomena, in Baëta 1968.

Barrett, D B 1971. Interdisciplinary theories of religion and African independency, in D B Barrett (ed): *African initiatives in religion.* Nairobi: East African Publ.

Barrett, D B (ed) 1982. *World Christian Encyclopedia*, OUP: Nairobi.

Chakanza, J C 1982. Towards an interpretation of Independent Churches in Malawi: *Africa Theological Journal* II/2 1982.

Chakanza, J C 1983. *An annotated list of Independent Churches in Malawi 1900-1981.* Dept of Religious Studies, Zomba: Chancellor College.

Chakanza, J C 1987. Religious independency in Southern Malawi, in Shank D A, 1987. *Ministry of missions to African Independent Churches.* Elkhart: Mennonite Board.

Chakanza, J C 1990. New religious movements in Malawi: a bibliographical review, in A F Walls and W R Shenk (eds) 1990. *Exploring New Religious Movements,* Elkhart: Mission Focus.

Daneel, M L 1971. *Old and new in Southern Shona Independent Churches,* Vol I. The Hague: Mouton.

Daneel, M L 1982. Fission dynamics in African Independent Churches in W S Vorster (ed): *Denominationalism - its sources and implications.* Pretoria: UNISA.

Elston, P 1972. A note on the Universities' Mission to Central Africa, in B Pachai (ed): *The early history of Malawi.* London: Longmans.

Gilliland, D S 1986. How "Christian" are African Independent Churches? *Missiology* 14/3 July 1986.

Hanekom, C 1975. *Krisis en Kultus: Geloofsopvattinge en seremonies binne 'n swart kerk.* Kaapstad: Academia.

Kachaje, D 1990. *Independent Pentecostal movements in Malawi 1970-1990.* Seminar on New Religious movements, Zomba: Chancellor College, (unpublished).

Kiernan, J P 1981. Themes and trends in the study of Black religion in Southern African. *Journal of Religion in Africa* 12/2 1981.

Kritzinger, J J 1988. *Die onvoltooide sendingtaak in Suid-Afrika.* Pretoria: ISWEN.

Martin, M-L 1975. *Kimbangu. An African prophet and his Church.* Grand Rapids: Eerdmans.

McCracken, J 1977. *Politics and Christianity in Malawi 1875-1940. The impact of the Livingstonia Mission in the Northern Province.* Cambridge: University Press.

Mwasi, Y Z 1933. *Black man's Church. My essential and paramount reasons for working independently.* Unpublished ms, Blantyre Synod. Copy at Zomba: Chancellor College.

Oosthuizen, G C 1969. Die konfrontasie met die onafhanklike kerklike bewegings in Afrika in die jongste Sendingteologie. *Sendingwetenskap vandag. 'n Terreinverkenning.* Lux Mundi I. Pretoria: N G Kerkboekhandel.

Oosthuizen, G C 1967. *The theology of a South African Messiah.* Leiden: Brill.

Pachai, B 1973. *Malawi: The history of the nation.* London: Longmans.

Parratt, J 1978. Y Z Mwasi and the origins of the Blackman's Church. *Journal of Religion in Africa* 9/3 1978.

Parratt, J 1979. Religious independency in Nyasaland - a typology of origins. *African Studies* 38/2 1979.

Parratt, J 1990. The Malikebu case, in Walls and Shenk 1990.

Pauw, C M 1980. *Mission and Church in Malawi. The history of the Nkhoma Synod of the Church of Central Africa, Presbyterian* 1889-1962. D.Th dissertation, University of Stellenbosch.

Pretorius, H L 1975. Waarom independentisme in Afrika? *Ned Geref Teologiese Tydskrif* 16/3 1975.

Ross, A 1968. *The origins and development of the Church of Scotland 1875-1926.* Ph.D thesis, University of Edinburgh.

Shank, D A (ed) 1987. *Ministry of missions to African Independent Churches.* Elkhart: Mennonite Board.

Shepperson, G 1968. Ethiopianism: Past and present, in Baëta 1968.

Shepperson, G & Price, T 1958. *Independent African. John Chilembwe and the origins, setting and significance of the Nyasaland Native Rising of 1915.* Edinburgh: University Press.

Sundkler, B G M 1961. *Bantu Prophets in South Africa,* 2nd ed. London: Oxford.

Taryor, L N K 1976. *Impact of the African Tradition on Christianity in Africa through the Independent Church Movements.* Ann Arbor: UMI

Taylor, J V & Lehmann, D 1961. *Christians of the Copperbelt.* London: SCM.

Thompson, J 1987. TEE and AIC. Challenges and problems of an interdenominational programme in Malawi, in Shank 1987.

Turner, H W 1967. A typology for African religious movements. *Journal of Religion in Africa* I(I) 1967.

Turner, H W 1979. *Bibliography of New Religious movements in Primal Societies* Vol 1: Black Africa. Boston: Hall.

Van Dijk, R A 1988. *Mini Swaggerts and Grahams: The young preachers of Malawi.* Unpublished ms Zomba: Chancellor College.

Walls, A F & Shenk, W R (eds) 1990. *Exploring New Religious Movements.* Elkhart: Mission Focus.

Wishlade, R L 1965. *Sectarianism in Southern Nyasaland.* London: Oxford.

CHAPTER 3
RELIGIOUS FREEDOM IN BOTSWANA: THE CASES OF THE ZION CHRISTIAN CHURCH AND THE NAZARETH CHURCH OF BOTSWANA

by
D. R. Boschman

Introduction

The history of the Christian church in Botswana is not widely known and the history of African Independent Churches in Botswana is even less known.[1] Tales from the early days of AICs in Botswana are fascinating and if asked, the majority of older, independent church leaders will tell many stories of persecution by tribal authorities and the British colonial administrators. These stories frequently end in 1966 with Botswana's independence from the United Kingdom and the proclamation of freedom of religion by Sir Seretse Khama, the first president of Botswana.

---

[1] Recently a number of theses have been written on Botswana independent church history. See Don Boschman, "The Conflict Between New Religious Movements and the State in the Bechuanaland Protectorate Prior to 1945" (Th.M. thesis, Harvard Divinity School, 1989), Rachel Friesen "A History of the Spiritual Healing Church in Botswana" (Th.M. thesis, Toronto School of Theology, 1990) and Obed Kealotswe, "Healing in the African Christian Churches in Botswana: A Comparative Study of Healing in Traditional and Modern Tswana Society, Old and New Testaments as Practiced by the Head Mountain of God Apostolic Church in Zion" (M.A. thesis, University of Botswana, 1985). For a bibliography guide to Botswana church history see James N. Amaze's recent monograph *Bibliography of Religion in Botswana* BOLESWA Occasional Papers on Theology and Religion 1:3 (Gaborone: University of Botswana, 1990).

In this paper I want to examine the beginnings of two Botswana independent churches, the Zion Christian Church and the Nazareth Church of Botswana. Both of these churches arrived in Botswana at about the same time and both were declared to be illegal organizations at about the same time (the late 1940s). But the results of those declarations were very different. In the end, members of the Zion Christian Church were expelled from their homes while the leader of the Nazaereth Church of Botswana was allowed to remain in his. This paper seeks to explore what circumstances led to these two different fates in order to give some nuance to Botswana's independent church stories of religious persecution.

This paper is based on two primary sources. The history of the Zion Christian Church has been reconstructed from government files held in the Botswana National Archives. The history of the Nazareth Church of Botswana comes from its founder, Bishop Motlapi Bolokwe. He was concerned that his church would forget its beginnings and thus, during the two years before his death in 1991, he frequently told me the story of his church. Douglas Dube, a friend of the bishop, has collected many of these stories and written them down.

Christian Beginnings in the Bechuanaland Protectorate

Botswana is a dry, sparsely populated country inhabited largely by SeTswana speaking people. After the Bechuanaland Protectorate was declared by the British in 1885, the BaTswana were governed by both their traditional authorities (dikgosi)[2] and a British administration based outside of the Protectorate in Mafeking. Up until the late 1920s, the British in theory

[2] Kgosi is a SeTswana word which is usually translated as "chief" or "king". Dikgosi is the plural form of the word.

advocated a system of "parallel rule" in which both the dikgosi and the British authorities operated their own, complementary administrations. In practice, however, these two administrations often overlapped and conflicted with each other. Beginning in the early 1930s and continuing into the period under discussion in this paper, the British tried to impose a system of indirect rule in which they attempted to rule the BaTswana via the dikgosi.

BaTswana first came into contact with Christianity through Wesleyan and Congregational missionaries. Both the London Missionary Society and the Wesleyan Missionary Society began sustained work with the BaTswana in the early 1820s in what is today the Republic of South Africa. David Livingstone of the London Missionary Society was the first missionary to live inside of Botswana's borders. In 1857 he was followed by missionaries of the German Lutheran Hermannsburg Missionary Society. During the next 25 years different missionary societies set up work among various SeTswana-speaking peoples. The largest of these, the London Missionary Society, worked amongst the BaTswana, BaNgwato, BaKwena and BaNgwaketse; the Hermannsburg Missionary Society worked amongst the BaLete and the Dutch Reformed Church worked amongst the BaKgatla.

It is important to note that initially all of the dikgosi sought to limit mission work in their territories to only one missionary society. In 1879, the most famous of all BaTswana dikgosi, Khama III, spelled out the rationale of this policy to several Roman Catholic missionaries:

> "If both religions, the Catholic and Protestant, are the same we obviously only need one of them. If they are different, then there will be constant strife between them and that could cause divisions among my subjects."[3]

---

[3] Charles Croonenberghs, July 24, 1879, in *Journey to Gubuluwayo; Letters of Frs. H. Depelchin and C. Croonenberghs, S.J.* Rhodesiana Reprint Library,

The Roman Catholic Church was not the only denomination which was at least initially refused permission to work in the Bechuanaland Protectorate. The Church of England was consistently refused permission to work in the larger reserves of the Protectorate and was only allowed to work amongst the BaKhurutshe and the BaKwena when their dikgosi became dissatisfied with the missionaries of the London Missionary Society. The Seventh Day Adventists were allowed to start medical work in Kanye in southern Botswana in 1924 only under condition that they did not do evangelistic work. Things were even more difficult for the Black American and Pentecostal churches. At the turn of the century, the African Methodist Episcopal Church was politely turned back by all but one kgosi in the Protectorate. During the 1920s, 1930s and 1940s missionaries of the African Evangelistic Band, Latter Rain Assemblies of South Africa and the Church of God in Christ were also refused permission to do mission work in the Bechuanaland Protectorate.[4]

The first African Independent Church in Botswana was the King Edward BaNgwaketse Free Church. Mothowagae Motlogelwa (circa 1850-1942) was a MoTswana evangelist who left the London Missionary Society in 1902 because of their refusal to ordain him. Initially only 44 church members followed him but within two years that number swelled to 750 as most of the congregation left the L.M.S. Eventually, however, the local kgosi turned hostile to this new church and Mothowagae was expelled from the BaNgwaketse capital. In his absence, people drifted back to the L.M.S. and when Mothogwagae died in 1942, his church died with him.

Volume 24. Translated by Moira Lloyd (Bulawayo: Books of Rhodesia Publishing Company, 1879), pages 131, 132.

[4] Don Boschman, "The Conflict Between New Religious Movements and the State in the Bechuanaland Protectotrate prior to 1945" (Th.M. thesis, Harvard Divinity School, 1989).

During the first one hundred years of Christianity in the Bechuanaland Protectorate (approximately 1850 to 1950)the relationship between the churches, the missionary societies, the tribal authorities and the British administration was confused and ill-defined. For example, while the missionary societies resented anything which resembled interference in their work, they did not hesitate to call upon the tribal authorities to build church buildings for them or upon the British administration to use its power in order to gain access to closed BaTswana towns and villages. Similarly, while in practice the British administration claimed to believe in freedom of religion, they did not hesitate to move against those religious groups which they believed to be working against the administration's and its allies' self-interest. The BaTswana dikgosi were, at least, consistent. They never claimed that there was any separation between church and state and they did not hesitate to exert what political authority they had in order to influence events in the local church.

The Zion Christian Church

A dispute over chieftainship lay at the heart of the creation of two new independent churches in the BaKgatla Reserve during the late 1930s. The trouble began in 1924 when Kgosi Linchwe died after an almost fifty year reign and was succeeded by his grandson, Molefi. Since Molefi was still a minor, he did not ascend to the throne until 1929.

Once on the throne, however, Molefi quickly alienated several important segments of BaKgatla society: his royal relatives, the Dutch Reformed Church and the British administration. In SeTswana culture, Molefi was not entitled to rule as a dictator, he was required to consult with his uncles. But because of his rivalry with his uncle Isang, Molefi turned to the commoners in his capital of Mochudi for friendship and advice. The Dutch Reformed Church was the second major power in the BaKgatla Reserve. The BaKgatla had brought this church with them when they left South Africa in 1871 and they had received their first D.R.C. missionary in 1877. Molefi, however, did not get along with the D.R.C. and its resident missionary Wilhelm Krige and in 1935 he asked his people to leave the church. They refused. Molefi then asked the Dutch Reformed Church to replace Krige. They also refused and Molefi had to back down. Perhaps most importantly, Molefi also alienated the British administration. As a result of this, Resident Commissioner Charles Rey deposed Molefi in 1936 and replaced him with Mmusi, his younger brother.

The majority of the BaKgatla did not like this move. One of the first signs of their displeasure came in Sikwane, a village about 50 km from Mochudi, the BaKgatla capital. In 1937 the MoTswana minister and most of the Dutch Reformed Church congregation there resigned from the D.R.C. and formed the Bakgatla Free Church. Mmusi acted quickly. Although his older brother Molefi had not yet formally joined the new church, it was clear from the Sikwane church council's letter of resignation that the new church had Molefi's support and that many of Molefi's supporters could be found in the church. Mmusi did the logical thing. He declared the new church to be illegal. Molefi's supporters responded by forming a second organization called Ipelegeng which started to collect money for Molefi's defence fund, but just as quickly Mmusi declared that this organization was also illegal.

Molefi's supporters then tried for a third time. In 1938 they sent four of their leaders to Morija, South Africa, to meet with Bishop Lekganyane of the Zion Christian Church. Bishop Lekganyane gave the delegation his blessing, made one of them a minister of the Z.C.C. and sent them back to the Protectorate. There, somewhat surprisingly, Mmusi allowed them to worship as a church but placed restrictions on their activities. Several prominent people, including Molefi's and Mmusi's mother and uncle, joined the new church.

Mmusi, however, was proving himself to be a weak ruler and soon the British began to seek ways to reinstate a rehabilitated Molefi to the throne. The deal the British finally offered Molefi was this: if he agreed to join the African Auxiliary Pioneer Corps and behave himself in the war in North Africa and Europe, Molefi would be reinstated as kgosi upon his return. Molefi agreed and headed off to war.

Three years later Molefi returned to the Bechuanaland Proteotorate a hero and was reinstated as kgosi. But his first decree as kgosi was completely unexpected: Molefi decreed that from then on, the Dutch Reformed Church would once again be the only church which would be permitted within his territory. The Zion Christian Church would have to be disbanded, its church building demolished and the green insignias of its members destroyed. Molefi had dramatically turned his back on the institution which had supported him during the previous eight years.

What brought about Molefi's change of heart? A number of factors probably contributed to it. Molefi's uncle had ruled the tribe for the previous three years and his mother had a great deal of authority: Molefi needed to assert his own control. Furthermore, in order to rule effectively Molefi needed the support of both the British administration and the Dutch Reformed Church, two institutions which were opposed to the Z.C.C. Thirdly, Wilhelm Krige, the white missionary to whom Molefi had objected so strongly in 1935, had been transferred and Molefi was impressed by his successor. Finally, Molefi probably did not want to share the loyalties of his subjects with anyone, and Bishop Lekganyane of the Z.C.C. now commanded some loyalty from Molefi's people. The Zion Christian Church had served its purpose in Molefi's attempt to regain the chieftainship, but now that he was kgosi, the church's existence was no longer politically desirable.

But if Molefi expected the Zion Christian Church to disappear quickly, he was to be disappointed. As the Bechuanaland Protectorate was quickly modernizing, many of the conservative teaching of the Z.C.C. appealed to those people who felt the disintegration of the old order. This included Molefi's own mother, Seingwaeng, and her husband, the ex-regent Bakgatla Pilane who decided to remain members of the Zion Christian Church. Molefi tolerated the Zion Christian Church's continued existence for a time, but towards the end of 1947 he decided to take decisive action against six of the leading families in the church. That one of these families included his own mother did not make any difference. Molefi had the families loaded on to the back of a truck and dumped off at the Gaborone railway station, 30 miles to the south. In the wake of such decisive action, 140 other members of the Z.C.C. publicly renounced their faith.

Although the British did not believe that the Z.C.C. posed any threat to Molefi's authority, they agreed with Molefi that any recalcitrant members of the Z.C.C. should be expelled from the BaKgatla Reserve. They were expelled and many of them moved to the nearby Gaborone railway station. In the winter weather, an old man, an old woman and two children died of exposure. By now, the Zion Christian Church in exile had swollen to about 150 people, including children.

For the next six years, the Z.C.C. exiles searched for a home. Finally, in 1953, the exiles were allowed to buy a farm of their own and eventually more than 3000 Z.C.C. members trekked 400 km north to Lentswe La Moriti in the Tuli Block. Today, the Z.C.C. still occupies the village of Lentswe La Moriti but it has also spread throughout Botswana, making it the largest African Independent Church in Botswana. It has, however, split into three factions. One is the St. Engenas Zion Christian Church which recognizes Joseph Lekganyane as its bishop and two others are simply called the Zion Christian Church, both of which recognize Barnabas Lekganyane as their bishop.

The Nazareth Church of Botswana

The BaLete are one of the smallest SeTswana-speaking groups in Botswana and they occupy the smallest district in Botswana. The BaLete were first evangelized by Lutheran missionaries of the Hermannsburg Missionary Society in the mid-nineteenth century. When the BaLete moved to their present capital of Ramotswa in south-eastern Botswana in 1875, the Lutheran Church came with them. The Christian community in Ramotswa was entirely Lutheran until the late 1920s when the Roman Catholic Church was able to buy a farm only 25 km from Ramotswa. Since the Lutheran mission was, at the time, doing very little medical work, the Catholics first made contact with the BaLete by conducting regular clinics. A few years later permission was granted to start a primary school in Ramotswa and in 1935 the Roman Catholic Church was allowed to start a mission in Ramotswa.

Motlapi Bolokwe, the founding bishop of the Nazareth Church of Botswana, was born in Ramotswa in 1906. Bolokwe's father, Sakaio Bolokwe, had been one of the earliest Lutheran converts in Ramotswa, and Motlapi Bolokwe was immediately baptized as an infant.

After being confirmed in the Lutheran Church in 1923, Bolokwe went to Kimberley to work in the mines. In 1930 he temporarily returned to Ramotswa to marry Kerebotse Rahube but he quickly moved back to South Africa to work. By 1933 Bolokwe had found work in the Johannesburg area and his wife was able to join him. Together they lived in Newclare, Sophiatown.

The Bolokwes' religious conversion began when Mrs Bolokwe became ill. Despite a very long hospitalization back in the Bechuanaland Protectorate, she did not recover and in 1939 Bolokwe took her to stay with a traditional doctor in Mmankgodi, a village near Ramotswa. Her condition improved, but her recovery was not complete.

During this time, Bolokwe continued to work in the Johannesburg area. There he heard about a prophetess named Rebecca Msimanga who lived in the Orange Free State but whose husband lived in Orlando. After receiving money for her transportation, Rebecca came to Johannesburg. Bolokwe sent for his wife, and together they went to see the prophetess.

The prophetess, Rebecca Msimanga, told the Bolokwes a number of things about their private life. She pointed out that when Motlapi Bolokwe's father had died some three or four years earlier, Motlapi had not slaughtered the required ritual ox. She went on to tell the Bolokwes that Motlapi would became a well-known minister and that Kerebotse had strong gifts of prayer and healing. Finally, she instructed them to be baptized by immersion and to fully give themselves to the Lord Jesus Christ. As a result, Motlapi and Kerebotse were soon baptized into the Nations Twelve Apostolic Nazareth Church in Zion. Its leader, Bishop Nyathi, was from Zimbabwe and most of the congregation was made up of BaTswana and Zimbabwean mine-workers.

While initially Kerebotse travelled with Rebecca Msimanga and developed her gifts of prayer and healing, she soon returned to Ramotswa to pray for people there. There large crowds were attracted by her gifts and many of her first patients eventually formed the nucleus of the yet to be formed Nazareth Church of Botswana.

Back in Johannesburg, tension was developing in the Nations Twelve Apostolic Nazareth Church in Zion. BaTswana church members felt that they were the only ones who were paying church tithes and that they were being taken advantage of by the Zimbabweans. Bolokwe even got up in church and related a dream in which his staff had swallowed Bishop Nyathi's staff. Eventually the BaTswana members withdrew from the church and formed the Nazareth

Apostolic Church in Zion.[5] Motlapi Bolokwe was elected bishop of this new church and Rantsootsoope Seasole was chosen to be one of his assistants.[6] On periodic visits back to Ramotswa, Bolokwe conducted prayer meetings there.

Soon after World War II Bishop Bolokwe returned to Ramotswa for good. It should be noted that this was the same time during which the Zion Christian Church was being expelled from Mochudi only 80 km to the north. In Ramotswa Z.C.C. members were also being expelled from BaLete tribal land. Today, it is alleged that at that time some Zionists were bound hand and foot and left in the sun until they renounced their faith. Others, it is said, were tied up and hung upside down from trees. It was this kind of climate into which Bishop Bolokwe returned.

For a while he was left alone. Then, on the evening of August 2, 1949 Bolokwe was arrested. The trial began on August 3, and lasted for three days. Even the Lutheran ministers joined in on his questioning. The principal charge against him was that he was operating an illegal church. Bolokwe's defence was that all of his churches were in South Africa and he did not have a congregation in Ramotswa. This did not satisfy the prosecution, and on August 9, 1949 Bolokwe was sentenced to a six month exile from Ramotswa.

The next day, Bolokwe returned on his own to the court to talk to the kgosi's representative. Bolokwe claimed that since he was being persecuted for

---

5 In the early 1970s, the name was changed to the Nazareth Church of Botswana.

6 Rantsootsoope Seasole was from Xhumo in north-central Botswana. He returned to the Bechuanaland Protectorate in 1947 and tried to settle in the Ramotswa area with Bishop Bolokwe but was denied permission to do so by the tribal authorities. He then returned to his home and began evangelizing the surrounding villages, becoming quite famous for his prophetic abilities. He founded many congregations and today many of the congregations of the Nazareth Church of Botswana trace their roots back to him.

praying in his own home village, the village elders were actually forbidding him to pray to God. The kogsi's representative did not have a good reply to this and eventually told Bolokwe to just go away. Bolokwe took this to mean that at least for then, the sentence against him was not going to be enforced.

Freedom of religion was granted to all BaTswana in 1966 and in the years which followed, the Nazareth Church of Botswana was able to construct a fairly impressive church building in Ramotswa. But the Ramotswa congregation remained small, consisting mainly of those whom Kerebotse Bolokwe had prayed for and healed. The church did, however, grow rapidly in other parts of Botswana and when the bishop died in June 1991, there were several thousand members of the Nazareth Church of Botswana scattered throughout Botswana.

Conclusions

The histories of the Zion Christian Church and the Nazareth Church of Botswana parallel each other in many ways. Both churches were founded in South Africa. Both churches came to Botswana during the 1930s and both churches were served expulsion orders in the late 1940s. But the fates of the two churches were quite different; the leaders of the Zion Christian Church were driven into exile from Mochudi while Bishop Bolokwe was allowed to remain in Ramotswa. What accounts for their two different fates?

The most significant factor must be the political roots of the Botswana branch of the Zion Christian Church. The Z.C.C. was initially formed to put political pressure on both the tribal authorities and the British administration and thus, even when its raison d'etre had ceased to be political, it was still perceived to be a political organization. The Nazareth Church of Botswana was not seen to have a political agenda. The size of the perceived political threat also had something to do with the sizes of the two churches. The Z.C.C. had between 150 and 300 members in Mochudi while Bishop Bolokwe claimed that a Nazareth Church congregation did not even exist in Ramotswa.

The BaKgatla and BaLete societies were also at two different stages of political development. The BaKgatla were an ecclesiastically homogeneous society under the Dutch Reformed Church and were ruled by a strong-willed kogsi. The BaLete had already become an ecclesiastically plural society with the arrival of the Roman Catholic Church in the 1930s and they were now ruled by a fairly weak kgosi. Thus, the existence of a new church in their midst did not threaten the established political order in the same way it did amongst the Bakgatla.

Finally, the personality of Bishop Motlapi Bolokwe was a factor, albeit not a major one. Although he was respected, he was stubborn. It should be noted that while Z.C.C. members in Ramotswa were physically mistreated and exiled, Bolokwe was not harmed in any way and he was allowed to defy the verdict of the court. The weak tribal authority of the BaLete tried to get rid of him, but he refused to be moved.

In conclusion, the roles of two other institutions should be noted. In both Mochudi and Ramotswa, clergy of the dominant missionary planted churches played a role in the persecution of the independent churches. The silence of the British administration is also significant. It does not appear that they played any role whatsoever in the persecution of independent churches in Ramotswa while in Mochudi they confined themselves to declarations about freedom of religion and the need for churches to stay out of politics.

BIBLIOGRAPHY

Boschman, Don 1989. The Conflict Between New Religious Movements and the State in the Bechuanaland Protectorate prior to 1945. Th.M thesis, Harvard Divinity School.

Botswana National Archives. Selected Files.

Croonenberghs, Charles and Depelchin H 1879. *Journey to Gubuluwayo; Letters of Frs. H. Depelchin and C. Croonenberghs, S J.* Rhodesiana Reprint Library, Volume 24. Translated by Moira Lloyd. Bulawayo: Books of Rhodesia Publishing company.

Morton, Fred and Ramsay, Jeff, ed 1987. *The Birth of Botswana A History of the Bechuanaland Protectorate from 1910 to 1966.* Gaborone: Longman Botswana.

CHAPTER 4
EVOLVING HARRIST VIEWS OF THE PROPHET HARRIS

by
David A. Shank

In 1922, Capt. Paul Marty, a French colonial administrator in the Côte d'Ivoire, wrote about the 1913-1914 impact of the Glebo (Liberian) Prophet William Wadé Harris, as "an almost unbelievable religious fact, which has turned upside down all the ideas about the black societies of the Coast, so primitive, so rustic; it will be - with our occupation and as a consequence of this occupation - the most considerable political and social event of ten centuries of history - past, present or future - of the maritime Ivory Coast".[1]

From Fresco in the west of the colony, an English commercial agent reported:

"Folks ... were sunk in debased superstition and fetich worship, and had been so for years. In three days, this prophet-fellow - I heard him preach myself - changed all that. Their fetiches were burnt, and what was an ordinary African coast village, steeped in superstition, became nominally a Christian town".[2]

From the eastern end of Harris's West African odyssey, Rev. Charles Armstrong, a Wesleyan missionary in the British colony of the Gold Coast, wrote on August 2, 1914 from the Methodist Mission in Axim to his district chairman:

---

1 Cf. Chapter one "L'impressionabilité des populations maritimes et lagunaires" in Marty, Paul, *Etudes sur l'Islam en Côte d'Ivoire*. Paris: Editions Ernest Leroux, 1922.

2 Mr Morgan, agent of Woodin and Co., cited by William J Platt, in *An African Prophet*. London: S.C.M. Press, 1934, p. 34.

"It appears that the "Prophet" Harris, whose doings you will have seen in the papers, has been here and through Apolonia, and as a result of his teaching hundreds have forsaken fetish-worship, burnt their fetish, and attend Christian services. On enquiry, I find that he preaches the doctrine of the Truine God, and Jesus as Saviour. He tells the people that in fetish-worship they are sinning against God. He attributes his success to the power of God and refuses to accept money. He does not attempt to found churches but directs all his converts (whom he baptises) to the Wesleyan or R.C. [Catholic] Church. "Go there," he says "and learn the commandments of God". [Gold Coast African Pastor, Rev] Bruce and his church leaders believed in the man, but I am not concerned about him of his teaching. The fact we have to bear in mind is that hundreds of people are seeking admission into our Church. They want to serve the living God and crowd in at our services. This is now the *normal* state of affairs and has been for three months. Whole villages have given up fetish and are asking for the gospel, chiefs are offering land for buildings and everywhere there is an awakening that we have prayed for, but scarcely expected perhaps".[3]

From Côte d'Ivoire, Monseigneur Moury, Vicar Apostolic, reported to the Apostolic See in his 1914 annual report:

"Space is lacking for exposing here the external means which the Divine Providence has used for the accomplishment of his merciful designs. I must thus limit myself to exposing the effects: a whole people having destroyed its fetishes, invades our churches en masse, requesting holy baptism".[4]

---

[3] Archives of Methodist Missionary Society (London), see dated Gold Coast files.

[4] Archives of the Société des Missions Africaines (Rome), 12/804.07: 28: 761, 1914.

From Liberia, the Cabinet of the Methodist Episcopal Church wrote in January 1916:

"One feature of our work in this field that should not be overlooked is the great revival movement among the natives, with which God has blessed us. But for this our membership could not have made the advancement it has. And yet we could not gather into the church all who professed conversion because we had not a sufficient number of missionaries to instruct and train them. Many however went into other churches and hence were not lost to Christianity. Literally thousands, largely young people, have been swept into the Kingdom of God, and were called upon to forsake their heathenism and superstitious practice for the sake of the Gospel. It is remarkable to what extent they do give up their old customs and habits, and many live such consistent Christian lives as to put some who were born and reared in Christian homes to shame".[5]

Father Ogé, head of the Roman Catholic mission in Liberia, wrote in 1920:

"Our missions are going ahead by leaps and bounds in the whole district. This is due not least - the Fathers themselves own it - to the former teaching of the famous Prophet Harris. Everywhere this extraordinary man passed, he ordered all the idols to be burned. His success was simply marvellous. Now the pagans, deprived of their old gods, stream to our churches and ask for religious instruction.... [He was] a boon to our mission".[6]

The prophet and his saga, the witnesses, the varied directions and spin-offs of the movement he generated, and the various interpretations of them by

---

5 *Liberian Conference Blue Book*. Monrovia, Liberia: College of West Africa Press, January, 1916. Cf.p. 7ff.

6 Cited in E M Hogan, *Catholic Missionaries and Liberia*. Cork: Cork University Press, 1981, p. 203.

Catholic and Methodist missionaries, colonial administrators, historians, socio-ethnologists, missiologists, scholars of religion comprise today a very extensive bibliography. Most of it is Western, by "outsiders looking in"; indeed scarcely fifteen percent of the items come from African writers, also with varied perspectives.[7] Only a half dozen very modest items are from those within: *L'Eglise Harriste*, the large religious community which emerged within the Côte d'Ivoire and bears the prophet's name.

Thus the appearance in Ivorian bookstores in 1988 of a substantial volume entitled *Le Prophète William Wadé Harris - Son message d'humilité et de progrès*[8], (henceforth *P-AHUI*) was a significant historical event for its author and the Harrist Church, as well as for the Ivorian religious scene generally. It is the first published history of the Prophet Harris and the beginnings of the Harrist Church as told by a lifetime Harrist and authorized officially by the Harrist Church's recognized and constituted leadership. Author W H Paul William Ahui is a second-generation Harrist, an intellectual, and a son of the "Supreme Preacher and Head" of the church. In the field of African religious studies this most important volume has not received the serious attention which it fully deserves. This is perhaps partly due to the French text in a field where French is often overshadowed or neglected, partly due to an ignorance of its existence because of poor promotion outside of the local context, and partly due to a lack of appreciation of its importance.

---

[7] For an overview of such materials up until 1980, see David A Shank, "The Prophet Harris: a historiographical and bibliographical survey," in *Journal of Religion in Africa* XIV, 2 (1983) 130-160, where more than 175 published items are surveyed.

[8] Ahui, W. H. Paul William *Le Prophète William Wadé Harris: Son message d'humilité et de progrès*. Abidjan: Les Nouvelles Editions Africaines, 1988, 348pp., bibliography, appendices, photos, index of names. Henceforth in the text: *P-AHUI.*

The present study is an effort to compensate for all of those factors in a manner that might also pay worthy personal tribute to Prof. Dr. G C Oosthuizen, who within the South African religious scene saw the need for valorizing so much that was overshadowed and neglected, for making available beyond its limited context such an important amount of material of inestimable value, and for enabling so many to appreciate its signal importance.

According to unwritten oral history[9], which becomes their written history in this volume, the village churches planted by the exceptional ministry of Prophet William Wadé Harris in 1914 in the French colony of Côte d'Ivoire, obeyed his renewed instructions to Ébrié chorister John Ahui in 1928, and started sending their youth to schools for training. This was in anticipation of their bringing back literate Bible instruction for their families, elders and peers. But, it is said, the net result of such obedience was to see their own youth become trained leaders either in the Roman Catholic or the British Wesleyan missions through the training received in the missionary schools. Indeed, as that oral history insists, it was often their own youth taught by the Roman or Methodist missionaries who most despised and looked down upon their prophet and the living symbols and traditions he had implanted in their illiterate communities.

With the coming of Ivorian political independence in 1960, and a national policy of heavy investment in modern education from primary to university levels, a new generation of Harrist youth was to emerge, trained in the French-speaking schools of the French secular tradition. That generation, which organized itself in 1975 in a National Committee of Harrist Youth, has since the mid-1980s carried a major role in the current national leadership of *L'Église Harriste*, under the wisdom, guidance and spiritual

---

[9] This is grounded in personal contacts in 1973 and 1974, research in 1978, and hundreds of visits, contacts and relationships during a decade of ministry in Côte d'Ivoire from 1979 to 1989. For a summary of those experience see David A Shank "The Work of the Group for Religious and Biblical Studies in West Africa" pp 15-77 in David A Shank (ed.) *Ministry of Missions to African Independent Churches*. Elkhart, IN: Mennonite Board of Missions, 1987.

authority of John Ahui, who had become the only fully recognized head of the scattered "Harrist" churches as they became organized nationally in the 1950s.

From its very beginning, among the preoccupations of this group of new intellectual leaders, two seemed to stand out: to justify the existence and validity of the Harrist Church from a historical and religious perspective in the face of its longtime Catholic and Protestant detractors; to present its own internal interpretation of the Prophet Harris in the face of dozens of "non-Harrist" understandings which have appeared over the years, as we have observed. These are prime preoccupations common to many of the African Independent Churches, due to their experiences 1) of rejection by Westernized mission-planted African churches, and 2) of having been studied and interpreted to the West, by Westerners, out-group specialists or Western-trained Africans. Here is to be found that deep inner compulsion, common to all human identities, to be one's own self-interpreter to other who think they know and understand, in order to provide a uniquely qualified and authorized self-understanding as a necessary prerequisite to dialogue.

The volume by W H Paul William Ahui[10] is the "mature fruit" of those preoccupations. It is so described in the preface written by President Tchotche Mel Félix of the National Committee of the Harrist Church of the Côte d'Ivoire under the authorization of "His Eminence JOHN AHUI, Supreme Preacher of the Harrist Church". This authorized work is all the more valuable and precious in that its author, as the son of the Supreme Head of the Church, was one of the early Harrist youth to profit from French university training, as a result of the prophet's instructions. This he did without abandoning the faith community of his family for a Catholic or

---

[10] The son of the Ébrié (from Abia-Gnambo, or Petit Bassam) Harrist preacher J Ahui, the author was the first son born (1936) after Ahui's return from his historic meeting with the Prophet Harris in December 1928, and was given the names of William, Wadé and Harris (thus the W and H) while the Christian name Paul was imposed by his tutor J Anoma in order to protect him from any repercussions of being a child from a Harrist family (cf. p 242 in *P-AHUI*).

Methodist (Protestant) identity[11] as would have been the case for practically all of his Ivorian university peers. But further, he has had unique access to the written and oral archives of his father, in addition to the family experiences during the formative period of the current Harrist institutions. Added to this is the personal experience as a "faithful Harrist" of some of the internal dialoguing and processing of history and tradition characteristic of communities enriched by cultural orality. Beyond this, however, he is the first Harrist to make serious use of published materials. In addition to the sources for general background, his bibliography lists some forty items, including some of the most recent French and North American research concerning William Wadé Harris.[12] His more intensive internal and external research has been carried on since 1984.

---

[11] Mr. Paul Ahui, French licentiate in natural sciences and a trained geologist, was in a second wave of Ivoirian students with scholarships for study in France. He was since 1981 - and at the time of writing and publication - the Director-General of PETROCI, the National Society for Petroleum Operations of Côte d'Ivoire, after having earlier been Director of Mines and Geology, and Coordinator of the United Nations Mining Project of the South-West [Côte d'Ivoire]. His spouse of French origin and his daughters, who supported him in this study, have shared a Roman Catholic religious identity.

[12] For the French: Bureau, René, "Le Prophète Harris et le harrisme", in *Annales de l'Université d'Abidjan (Institute d'Ethnosociologie)* Série F(3) 1971-73; Holas, Bohumil Théophile, *Le séparatisme religieux en Afrique Noire: l'exemple de la Côte d'Ivoire*. Presses Universitaires de France, 1965; Piault, Colette (ed.), *Prophétisme et thérapeutique*. Paris: Hermann, 1975.
For the North American: Haliburton, Gordon, *The Prophet Harris*. New York: Oxford University Press, 1973 (Abridged). Ahui does not use its French translation by Marie-Noëlle Faure (Abidjan: Les Nouvelles Editions Africaines, 143pp.), nor the more complete edition published by Longmans in 1971. Three important articles by Afro-American Sheila Walker are listed, but he does not refer to the rewrite of her significant 1976 University of Chigago doctoral thesis: Walker, Sheila, *The Religious Revolution in the Ivory Coast; The Prophet Harris and His Church*. Chapel Hill, N.C.: University of North Carolina Press, 1983. And finally, Shank, David A, *A Prophet of modern times. The thought of William Wadé Harris, West African Precursor of the reign of Christ*, 3 vols. University of Aberdeen (Scotland), Ph.D dissertation, 1980.

Reference in the preface to the "mature fruit" of preoccupation and study suggests that this is not the first of Harrist efforts to properly and correctly portray for themselves and others a consensus of self-understanding and interpretation. Indeed, several such attempts have appeared over the years, and it is perhaps not inappropriate to bring them into view as precursors to the work by P. Ahui, particularly since they are not listed in his bibliography nor referred to in this work.

Certainly the first effort to officialize such self-understanding by publishing was the *Premier livret d' éducation religieuse à l'usage des Missions Harristes* of 15 pages dating from 1956. The origin of publication is indicated as "Programme de l'Episcopat, Petit Bassam". Petit Bassam is indeed the "Mother Church" where John Ahui has presided since 1935. But the expression "Programme de l'Episcopat" is a language foreign to the Harrist movement itself, since "episcopal/bishop" is not a part of the congregational structures traditionally attributed to Harris himself. Quite clearly, twenty years after the beginning of his ministry[13] beyond Petit Bassam, John Ahui was being recognized generally, because of his time with Harris in 1928, as one with an authoritative power and word. But "Episcopat" has decidedly non-Harrist overtones. Several suggestions might explain: an admiration for Roman Catholic structures to be copied, an effort by Harrists to indicate that a valid parallel to Roman Catholicism existed within the Harrist movement, or even Catholics-become-Harrist, bringing into a counter-movement the current status-language of the best-recognized Western religious institution. John Ahui himself was illiterate, as were the other elders and preachers at that time. Thus the writing was necessarily done by (rare) intellectual members, or sympathetic friends.

---

[13] Reported in *P-AHUI* on p.210ff. in the first person, as told to him by his father.

At the time of our own first contacts with Harrists in 1973, I was presented with an old, hard-worn copy of the *Premier Livret* and was told it was the "Harrist catechism", when I asked for an understanding of Harrist teachings. Indeed it is written in a typical Roman Catholic catechetical style of question and response. It was then no longer in current popular usage among Harrists, and I was to discover in ensuing years that many Harrists had never seen it. But it clearly was still an important and significant reference work, wherever it was yet to be found. And for the older people, it was evidence that they too had their "intellectuals" who "*connait livre*" [know book=are literate], and had also given the Harrists their own "book". Quite indicative of the understandings promoted by this official document are the two following sections of "question/response".

Q: How many religions are there?

A: God sent to each people a prophet to reveal its religion to it. There are therefore as many religions as there are peoples who have received a prophet.

Q: Which religion was revealed to Africans?

A: It is the religion taught by the Prophet Harris and continued by his successors. (p3)

Q: Must one help and love also his enemies?

A: If someone says he is your enemy, even if he does you wrong, God alone has the right to punish him. As for you, you must continue to love and help him as a brother. No-one has the right to punish his neighbour.

Q: And if the enemy strikes us or insults us?

A: The prophet suffered humiliation and violence from the French authorities; never did he return a blow. On every occasion you must show yourself superior to your enemy by the forgiveness which you grant him even in spite of himself. (p12)

At the present time, it is nigh impossible to determine to what extent these were either self-understandings of the movement written down by helpful writers, or rather understandings of "intellectuals" in the know, that were hopeful of projecting them upon the whole movement through literary development. Efforts over the years to identify the writers were never conclusive, although over and over again, "Dr. Claver", "Bruno", "Bruno Claver" was the one name that recurred. The leading Ivorian psychiatrist in the 1970s, Dr Bruno Claver was - in the words of Harrist leaders - "our intellectual", and had been assigned to us as visitors in 1973 when we had our first contacts with the Harrist leaders.[14] Subsequent conversations with him always downplayed the significance of the *Premier Livret*, which he attributed to "a Frenchman, a certain Martin, who had a lot of sympathies for Harrists, and tried to give his resume of what they stood for". Our every effort to trace and learn more about "Monsieur Martin" failed; several younger Harrists even suggested that it was perhaps a pseudonym for a paternity denied.

Be that as it may, it is clear that those "intellectuals" who developed the official 1956 document were clearly distancing themselves from a Christian understanding of the movement, and broadly identifying it with the then current French vogue in African sociology which saw African religion as pre-literate political expression, and prophetic movements as colonized societies in transition

[14] He was in an early wave of Ivoirian students to Paris, and was apparently "caught" because of his alleged marxist sympathies, when Houphouet-Boigny rejected solidarity with the French Communist Party support that he had used to get the political movement for independence under way. A Catholic also, he discovered that the Roman Church in Europe was undercutting the movement for Ivoirian independence which he was leading as a student in Paris; he henceforth identified "as an intellectual" with the neo-Harrist movement that was emerging particularly among the Ébrié Harrists in the area around the political centre of Abidjan.

toward political independence.[15] This was no doubt also one important tendency in the developing Neo-Harrist revival among the Ébrié around Abidjan in the late 1940s and the 1950s, where Harris was perceived as wholly unrelated to the foreign,colonial religions of Catholicism and Methodism. He was rather an African uniquely for the Africans (as over against the French, or "whites"), a prophet of African autonomy and progress. Harris's own full self-understanding is not explored in the document. But what is remarkable about the document's projection of the movement and of Harris himself at that moment in its history is, of course, the clear Christ-like rejection of violence and vengeance, albeit with Prophet Harris as the model. It was a most significant grass-roots contribution to a much larger African anti-colonial, independence movement, with a major impact upon the popular mindset in what was to be largely a non-violent movement in the Côte d'Ivoire.

A second publication unlisted in P. Ahui's bibliography is Boyé Aké, Alphonse, *Le Harriste face à sa Religion* published in 1980 personally by the

---

[15] It is well reflected in George Balandier's *Sociologie Actuelle de l'Afrique Noire* (Paris: P.U.F., 1955), a major academic contribution to African political independence. It published his earlier courses, then so closely followed by the African intelligentsia in Paris. Here the Harrist movement in West Africa was perceived to be one of the "messianic movements" carrying significant political overtones. In those times of pre-independence, the intense activity going on within such religious movements did in fact oppose Western religion (=Roman Catholic and Methodist in the Côte d'Ivoire, seen an handmaidens of the repressive colonial regime) and were unquestionably also political in nature. It is in fact the time of development of a "second wave" of Harrism - what Amos Djoro calls "neo-Harrism" (cf. "Les églises Harristes et le nationalisme ivoirien," in *Le mois en Afrique* 5; May 1966, 26-47), and is confirmed by John Ahui in *P-AHUI* (p. 252). Ironically, it is Balandier's "Western" interpretation, done in the light of the "neo-Harrist" movement, that is projected back ideologically upon the whole Harrist movement including Harris himself, by the editors of the text for the neo-Harrist leaders; by so doing they increased considerably its local appeal and potential.

author, an Apostle in the Ébrié Anono congregation, and an employee in a printing house. Coming a quarter of a century after the "first" official publication, the brief pamphlet of 37 pages reflected the author's personal concern for the great diversity of practice which he observed among Harrists, as he paid visits to numerous congregations. He was interested in making his "own contribution in the search for a uniformisation of the structures of Harrism". Here the Prophet Harris is perceived as a mysterious personage and a powerful liberator who delivered the Ivorian masses from ignorance; his "arrival transformed events very much, [for] our grandparents now had their eyes and ears opened". The mission of 1913 and 1914 is seen to be that of "teaching the good news of God and of Christ, teaching the word of God and his commandments, and recommending to pray to God and to observe his commandments, so that they would one day come out of darkness".[16]

In a quarter of a century there is a shift in an official understanding of Harris's mission from one of political liberation to a personal one of spiritual liberation through the "good news of God and of Christ". Now there is accent on religious integrity and authenticity, with a serious call to reform and consistency of life, based on a deep faith in God. "The Harrist must be open, straightforward, peaceful, and especially must have faith in God by scrupulously respecting his commandments". A description of structures and patterns of worship is included. But there is also a strong accent on Harris having transmitted his power to "Jonas Ahui" for the continuation of his work. Other self-designated "successors" have turned many African faithful believers away from their religion. Among these, the most important was the "clever Methodist Pierre Benoît who was able to fake with remarkable ingenuity a photograph of

[16] A more thorough review can be found in David A Shank, "A Window into the Harrist Church" in *Mission Focus*, 9(3), September 1981, 55-59.

the Prophet Harris shaking his hand". Here then is found a preoccupation with defending the Harrist institution as over against the competing claims of others for the Prophet's succession.

Although this was not an "official" publication, it did emerge from one of the more progressive congregations whose "Superior Preacher" was an intellectual, working as court clerk in his secular calling. We discovered that the brochure was known in numerous churches we visited, and for its time must have had considerable internal influence as an almost unique Harrist publication. But it must also be pointed out that the author had new resources for understanding the prophet. The French-trained Holas had gathered a considerable bibliography on Harris in *Le Séparatisme Religieux en Afrique Noire* (Paris: P.U.F.) from 1965. The French René Bureau had further opened up the religious perspective in an important article from 1971 that used many sources dating back to the early time of Harris's passing through the Côte d' Ivoire.[17] An Ivorian historian, Professor Christophe Wondji of the University of Abidjan had authored in 1977 a 96-page small paperback on the Prophet Harris in the popular series "*Grandes Figures Africaines*". *Le Prophète Harris: le Christ Noir des Lagunes* (Paris: ABC and Abidjan/Dakar: NEA) was die 33rd volume in the series edited by the distinguished historian Ibrahima Baba Kaké. Wondji asked openly whether Harris was not a messenger of Christ fulfilling the Great Commission, and if so, why he was rejected by Catholic and Protestant missions. Yet he recognized that the prophet was also considered to be an African nationalist seeking to give to his brethren the means of cultural resistance to a foreign domination. That ambiguity is accented in the subtitle "The Black Christ of the Lagunes" which was Wondji's perception of the then current Harrist understandings of the prophet,

---

[17] René Bureau, Le Prophète Harris et le religion harriste (Côte d'Ivoire) *Annales de l'Université d'Abidjan, Institute d'Ethnosociologie,* Série F (3), 1971, 31-96.

this despite the historical evidence generally unknown to the Harrists which accented Harris's Christian orientation and message. As a historian, he had at his disposal at least eight significant sources dating from before 1936, when John Ahui was to begin his inter-village ministry beyond that of Petit Bassam. But he also had at his disposal Piault (1975), Haliburton (1971), and Bureau referred to above. Though there are numerous errors in dates and understandings of Harris's pre-prophetic life, there are also numerous stories from oral tradition of the Côte d'Ivoire that found their way into the text. But the most significant is that Harris's own self-understanding as one sent by Christ to preach the message of Christ becomes clear from the early documentary sources. This perception also shines through the Boyé Aké study. This suggests that the story as told by Wondji had a significant influence in the shift that we have seen, as it appeared among the growing numbers of literate youth, of which Boyé Aké was typical.

The continued preoccupation about the validity of John Ahui's succession to the Prophet appears again in 1983, in another 35-page magazine-like publication entitled *Jonas Ahui, Chef Suprême de la Religion Harriste, Successeur du Prophète William Wadé Harris*. Written and published personally at his own expense by Kouassi Boussou Gôh Benoît, who was then Superior Preacher of Dougbafla (Oumé), it was prepared nevertheless with the partial collaboration of Supreme Preacher John Ahui, who for the first time was to be gratified with the publication of the story of his visit to the Prophet Harris between December 5 to 21, 1928 as told by himself.[18] It is a major oral document, printed extensively for the first time, and which gives detail upon detail which would

---

[18] Taken from a tape recording made early in 1976 by M Robert Arnaut of *Radio-Télévision Francaise* and made available by the Documentation and Resource Centre of the Group for Religious and Biblical Studies in West Africa at Abidjan. Arnaut had earlier published a part of the story in *L'Afrique du Jour et de la Nuit*. Paris: Presses de la Cité, 1976 pp. 72f.

seem to preclude any doubt concerning its authenticity. Here is found the personal experience of Harris's charge to Ahui, his giving the cross-cane and the Bible, the transfer of power, the prayer of consecration, the Christ apparition.

Author Gôh cites his personal motivation for publishing, namely the exhortations to him from His Eminence John Ahui at the time of his consecration as Harrist preacher: "If the Harrist religion were scarcely to evolve, it would be the fault of the intellectuals. Hence the challenge publicly given to Ivoirian youth". It was, however, a private initiative, a personal publication, financed with personal funds, reflecting the personal enthusiasm of the author for the Harrist Church. In 1984 the same author, who now designated himself with four major official functions in the Harrist Church, was to produce a very highly detailed description of Harrist funeral practices, with a somewhat Roman Catholic ethos, but also with citation of Scripture upon Scripture for justification of one or other practice.[19] Indeed, the author was himself a relatively recent convert-newcomer to the Harrist Church and clergy, coming from a second-generation Roman Catholic family, and was in his secular calling a "Technical Counsellor" to the Minister of Labour. Zealous in his ministry, as an intellectual he needed to justify before his friends and colleagues his decision to identify with such a marginal group. Some of the personal interpretations and explanations are indeed in a language reflecting a Roman Catholic religious orientation.

In *Jonas Ahui* the presentation of Jonas Ahui's story in the context of "L'Affaire du Pasteur Benoît - Un Loup à la Peau d'Agneau" [A Wolf with a Sheepskin] (p.8), indicates the importance of the Harrist community's long term

---

[19] Kouassi Boussou Gôh Benôit, Prédicateur Supérieur Harriste, Sécrétaire général adjoint du Comité National Harriste, Membre de la Commission chargée de la Formation des Prédicateurs, Membre de la Commission chargée de la liturgie *Les Funérailles Harristes* [Abidjan, published by author], Presses de l'Imprimerie Nationale, 1984, 71pp., 49 photos. The description in detail and practice suggests an ethos similar to the biblical book of Leviticus.

interpretation of the British Wesleyan Mission's achievements in 1924. What the latter - with all of the Western missionary community - saw as missionary responsibility in 1924 is still today perceived by the former as Harris being duped by Methodist "religious covetousness" in which occurred, with earlier Catholic recuperation of Harrists, "the dismemberment, if not the upset of 90% of the Harrist edifice". From such a perspective it was obviously most fundamental to justify historically the fact that Jonas Ahui - not Pasteur Benoît and the Methodist Mission - is indeed the successor to the Prophet Harris, and the unique head of the church. This is history perceived and told from its "bottom side", rather than that of the "top side", colonially authorized and established, mainline, Westernized church.

Another seven pages are given to explaining the calling and function of the Harrist healer, Albert Atcho of Brègbo (deceased in 1990). In the very personal view of the author, Atcho's prophetic healing gives continuity to a parallel, yet different, dimension of Harris's ministry; indeed, for him Ahui and Atcho together fulfil the mission of Harris. In addition to contemporary colour photographs of Ahui, Atcho, and of course, the author, there are the historic photos of Benoît's meeting with Harris in 1926, and Ahui's investiture visit in 1928.

Such were the Harrist publications which preceded the substantial volume by Paul Ahui. And it was his wife's acquisition in 1984 of a work - apparently *Jonas Ahui* (21), a book that pretended to speak so authoritatively about his own father - which ultimately provoked the research and study by Paul Ahui. "I devoured the book and, cut to the quick, I understood the requests of the Supreme Preacher and of the Harrist community" (*P-AHUI* p.9)".

In other words, despite the good intentions of a recent Catholic-become-Harrist, the work was highly inadequate, and it was time for a competent, authentic Harrist to tell the whole story correctly. Indeed, author Gôh's motives were never always clear to many other leaders; by 1987 he was under much suspicion and by 1989 had been defrocked.

P-AHUI is first of all a most important work since it offers to the French, the Ivorian, and above all the Harrist reader, reproductions of a number of very significant documents related to the history of the Côte d'Ivoire and the early days of the Harrist movement. These have been known and interpreted by specialist and/or anglophone readers, but *P-AHUI* puts them at the disposal of the general francophone public. They are as follows:

1) Nine letters from April 1914 illustrating the Colonial Administration's need to deal with the petty quarrels between Catholic and Wesleyan missionaries in the Côte d'Ivoire - an insight into the confessional religious strife which characterized the colony at the time when the Prophet Harris exercised his own African mission with tolerance of all Christian confessions.

2) A letter, dated September 1926, from the head of the Catholic Mission at Bingerville, indicating the intention of establishing Catholic chapels and schools in four Ébrié villages, and reminding that prior to a decree of February 1922, they had already done so in a dozen other Ébrié villages - indicative of the way in which the Roman Catholic Church was exploiting the Harrist legacy in its own interests, and building upon it.

3) A letter from the Catholic mission at Bingerville, asking for authorization to take up a collection "without distinction of religious opinion, because missionaries do not only fulfil a religious work, but also a patriotic work which ought to interest all the good French" - illustrating the connivance of Catholic Mission and French patriotism, implied in the original contract between colony and missions.

4) [A photocopy of] Copy of Methodist Missionary Pierre Benoît's handwritten French "Harris Testament" of September 25, 1926, with signatures and fingerprints of Victor Tano [P Benoît's translator, and one of Harris's translators

in Côte d'Ivoire], M Brewey, Mrs Neal {Harris's daughter Grace], Pasteur P Benoît, W W Harris.

5) [A photocopy of four pages of] Messages from Prophet Harris, December 1928.

6) Announcement of the death ["Aply {=April} 23th 1929"] of the Prophet by G Nelson.

7) Four handwritten texts from June 1932, by an occasional anglophone teacher in the Petit Bassam Harrist Church.

8) Photographs:

(a) In annex:

"The Prophet W W Harris with his daughter and his grandchildren, as well as the pastor Pierre Benoît".

"The meeting of Prophet W W Harris and of pastor P Benoît of the Wesleyan Methodist Church, at Cape Palma, the 25 September 1926".

"The Prophet Harris as seen at Axim in 1914".

"Ready for the day work" [picture from Harris's ministry in Gold Coast].

"baptizing a fetish princess" [also from Gold Coast]

(b) In text: 11 other photos, including the photographic proof of John Ahui's visit with Harris in December 1928 (p.196); five maps.

9) Very significantly, the original text in English of Pastor Martin Luther King's "I have a dream".

10) The background and reaction to the coming of Harris as experienced by the village of Abia-Gnambo (Petit Bassam), pp.125-154.

11) The full story of John Ahui's visit to Harris, reported in first person present tense, pp. 181-200, with some differences from the text in Gôh Benoît's publication reported above.

12) John Ahui's personal story of his beginning of the ministry which Harris committed to him, pp.209-214.

13) Stories of "power confrontation" from John Ahui's ministry of evangelisation, with collaboration of Peter Loba (the older) and Nowa Tchê-Tchê in the years of the late 1930s and the 1940s, pp.217-236.

Secondly, for the first time there is a "Harrist" interpretation of these documents, showing their significance in the light of the Harrist Church's experiences and understandings. That interpretation is the story of the survival, renewal and structured continuity of an African religious movement, community, and tradition which - for the colonial authority, the colonial Roman Catholic and British Wesleyan missions, their African churches and other Westernized Christian bodies---was destined to disappear with their help and proselytism, because of its troubling political, religious and social inappropriateness. But it is also the story, within changing contexts, of the planting and reaping of the seeds of liberation - from slavery of Blacks in the United States to an independent Liberia; from subservience of the Glebo under Americo-Liberian political and religious control in Maryland of Liberia to a God-given human dignity; from the "fetichism" of traditional Africa to a Bible-related, African humanism; from the hostility and conflict of "tribalism" to humble, non-violent human interaction in an intra-ethnic community of faith and tradition; from the ignorance of the "other" and blind submission within an oral, ethnic, ancestral past and heritage to modern education in literacy, in the Christian Scriptures and faith, in thought and understanding of the world, in progress from the sciences...; from forced labour and colonial oppression to paid labour and political independence.

Thirdly, one is impressed by the detailed exposition in French of the creation of the Liberian political context and background from which the Prophet Harris emerged on the scene of history. One might regret here the absence of the manner in which the pre-prophetic Harris related to the political and religious forces present: Rev. Lowrie, not as an Americo-Liberian, but as a freed

Glebo slave converted and trained in Sierra Leone; the role of the Americo-Liberian Episcopalian Bishop Samuel Ferguson in imposing the oppressive establishment's cultural, political, and religious model on Glebo converts; the criticism of Western Christian missions and the recognition of values of Africanity, and promotion of a pan-African Christianity (and "independency") by the distinguished Dr. Edward Blyden; the disruptive religious "independency" promoted also by the highly apocalyptic expectations of "Russellism"'s *Watchtower* in the emerging movement of the Jehovah's Witnesses; the major shifts of Harris from "assimilated Liberian" to "emancipated native", and thence to Christian prophet as "forerunner of Christ's Kingdom".

Indeed, to the extent that the perception given in the volume represents that of the first generation of Harrist intellectuals, one can say that they would appear to ignore completely the biblical apocalyptic eschatology which so clearly provided the dynamic for Harris's unequalled mission. On the other hand, there is a clear and open insistence upon the Christian character and thrust of Harris's ministry. The author particularly accents a lifestyle of African humanism featuring forgiveness and love; peace and non-violence with fraternity among differing ethnic groups; tolerance of the traditional family and between differing Christian and prophetic movements; humility as he perceived it transmitted from Harris to his father-head Preacher. "The principal goal of his mission was the revelation of the Christ Saviour to the Africans... If this aspect did not enough catch the attention of the first Harrists, the future of Harrism requires a greater awareness of this fact which cannot be disassociated from the vision of the Prophet. It is why the Harrist liturgy ought to give a greater place to reading of the Bible and commentaries of the teachings of Jesus Christ (p.261). Seen in the perspective of the 1956 official *Premier Livret* this represents indeed a major christological shift... in intention. This is a most significant thought development for indeed the neo-Harrism among the Ébrié and the Attié, for example, still had in the late 1980s a strong continuing tradition which said that "Jesus Christ came for the Whites, the Prophet Harris came for the Africans".

Finally, the author's open recognition of this early and continuing christological vacuum within the movement might have allowed for some modification of the interpretation of the Pierre Benoît visit and "testament" of 1926 by today's Harrist intellectuals. But here Paul Ahui follows the oral tradition's hard line (pp. 171-276, 205-207) which accuses the Methodist Mission of bad faith, of intentional duping of the prophet, in the sole interests of church politics. Despite the evidence, he even doubts the authenticity of Harris's signature on the famous "testament", which was in any case not an unadulterated Harris document, according to the evidence. It is a tradition which is understandable, given the historical developments, but which appears to refuse *post facto* to Harris the freedom to have changed his mind between 1926 and 1928 on the basis of some of those developments.

This tradition seems also to ignore the major christological concern of the Methodist Mission after the mid-1920s, in the face of the then emerging and divergent "prophetisms" totally lacking in the biblical and christological orientation of William Wade Harris. Sixty years later, the Harrist Christ-oriented author now shares that concern (p. 265) in the current context. But his positive report of post-Harris developments in Petit Bassam under the influence of "preacher Toh" (p. 169f) reflects the older people's appreciation and gives some indication of the way in which the Harrist movement developed without the later provocative parallel presence of the "Christian" Methodists and Roman Catholics[20]. The "prophet Dôh" was indeed a semi-literate syncretistic Kruman

---

[20] However, Gordon Mackay Haliburton, *Le Prophète Harris* (Abidjan: Nouvelles Editions Africaines, 1984, 143pp) also appeared that year, yet *P-AHUI* does not list it in the bibliography. The translation had been done by Marie-Noëlle Faure, subsidized by the Mennonite Board of Missions (USA),

whose thought and practices were similar to those of Harris in appearance only[21].

It is the official character of this long-time Harrist view of Pierre Benoît's 1926 visit, reinforced by the published account of John Ahui's personal understanding (p.195) and report of Harris's own self-punishment, which continues to perpetuate a deep gap between today's Methodists and Harrists in Côte d'Ivoire. *P-AHUI* has thus also provoked an Ivorian Methodist rejoinder: E.[rnest] Amos Djoro *Harris et la Chrétienté en Côte d'Ivoire* (Abidjan: Nouvelless Editions Africaines, 1989 60pp), based on research completed at the Sorbonne in 1956. With a preface by long-time [Methodist] Minister of State Mathieu Ekra, the modest volume also deepens that inherited gap by denying that Harris's donation of his cross and Bible to John Ahui in December 1928 had anything to do with a formal sacerdotal ordination. It gives the Methodist "missionary" perspective: Harris was a Methodist, expected the churches in Côte d'Ivoire to be Methodist, gave such a testament to P Benoît, and died a Methodist. (In fact, in 1888 he had left the Methodist Church for the American Episcopal Church mission in which he raised his family). He had promised his baptized converts

---

at the request in 1973 of National Committee President Mathieu Keingba to Mennonite emissary Dr. Marlin Miller, who had earlier given Harrist leaders a copy of Haliburton's unabridged edition for information, verification and consultation. At the time President Keingba was determined to publish the volume at Harrist expense; in fact resources were not forthcoming, and NEA agreed to publish the text as a contribution to Ivoirian history and culture.

[21] This can be observed among the ébrié in *Cantique Harris Ébrié* (Wêlê Nan-Mê Rho). Abidjan: Les Presses de L'Imprimerie Secco, [1984?]. The influence and development among the Dida Harrists has been traced by James Krabill in *The Hymnody of the Harrist Church among the Dida of South-central Ivory Coast (1913-1949): An Historico-Religious Study*. Ph.D thesis, University of Birmingham (England), 2 vols., 1989. The hymns themselves, compiled by Beugré, Kobli Alphonse and Krabill have been published in 1988 in four numbers of *Chants Harristes Dida*, by the Summer Institute of Linguistics at Abidjan.

that white missionaries would come after him with the Bible and teach them, and the British Wesleyan Mission was indeed the fulfilment of that promise. Hence, in this perspective, the Ivorian Harrist Church was never a part of Harris's intention, and is the only place where such an anomaly grew out of his work (thus ignoring "The Church of the Twelve Apostles" in Ghana).

Even though Amos Djoro recognizes that the Methodist Mission broke with Harris's methods of approach by seeking to apply to "Ivorians scarcely delivered from paganism, all of the rigor of Methodist laws and disciplines conceived for the subjects of His Gracious Majesty" (p.49), he does not draw the conclusion which is imposed. In his defence of the authenticity of the 1926 "testament", it does not occur to him that the consequent Methodist rejection of Harris's insistent tolerance of polygamy, and the imposition - in Harris's name - of what appeared to be a church tax, would meet with Harris's hot indignation and regret, as well as a feeling of having been betrayed. Indeed, Benoît's own notes indicate that Harris had given clear contrary instructions to him on both these questions.[22]

In fact, the document which John Ahui brought back from Harris in 1928 reflects this clearly. "In the name of God" it approves of plural wives, warns of the "false prophets the one who receive money on". It further announces that France will be "terribly scourged" because of their treatment of the prophet. Two years earlier this message, too, had been given to P Benoît for the French colonial governor; it was never delivered. And of course, the British missionaries never again consulted Harris to see how satisfied he was with their

---

22 See Shank, *op cit*, Note 12 above: Vol. 3, Appendix XLVII Benoît, pp. 1069-1130. The original notes are in the Methodist Missionary Society's archives (West Africa/Ivory Coast) in the School for Oriental and African Studies in London. They were not indexed when discovered in 1977. A photocopy is in the holdings of the Documentation and Research Centre of the Group for Religious and Biblical Studies in Abidjan, Côte d'Ivoire.

self-perceived fulfilment of his mission. The Methodist intellectual spokesman Ernest Amos Djoro, like his Harrist friend and peer Paul William Ahui, has apparently also refused to the Spirit-led prophet the freedom to have changed his mind between 1926 and 1928.

The *P-AHUI* volume understandably does not foresee a rapid ecumenical meeting between Harrism and the other Christian churches (p.276), although this reviewer perceives that the potential for genuine dialogue is greater than at any previous time, since with this volume many of the necessary elements are now in the public domain. And further, the author - in the spirit of Harris as developed by Martin Luther King and with the eschatology of a Teilhard de Chardin - is convinced of an eventual and necessary convergence...

It is a remarkable volume, not because it tells us something new about Harris, for it does not, but because of what it tells us about the Harrists and their movement from traditional Africa into the future as "protestant Christians with an African connotation" (p. 273).

CHAPTER 5

SHIBE AND THE FORMATION OF THE ZULU CONGREGATIONAL CHURCH

by

A. W. Z. Kuzwayo

In the historiography of the African Independent Church Simungu Bafazini Shibe has often been depicted as a villain who had motives other than ecclesiastical when he and Fokoti Tyoyo[1] broke away from the American Zulu Mission and founded the Zulu Congregational Church on the 14th of February 1898. Shibe and Tyoyo were Approved Preachers at Table Mountain and at Elandsfontein in Johannesburg, respectively, under the American Zulu Mission and, therefore, could be considered for ordination as pastors. By the middle of the year Tyoyo had died and the leadership of the new church in that area was taken by an ordained pastor, the Rev Sunguza Nyuswa, specially recruited by the new church from Mthwalume in Natal.

Shibe's and Tyoyo's secession was not the first one in the history of Christian missions in South Africa. Indeed, secessions of Christian converts from mission societies and the formation of independent churches under African leadership characterised colonial Africa from the second half of the nineteenth century well into the twentieth. When they occurred, these secessions had been

[1]Lester Ernest Switzer, *The problems of an African Mission in a White-dominated, multi-racial society: the American Zulu Mission in South Africa, 1885-1900*. D Phil (Natal), 1971, p. 414, Note 50, suggests that the name could have been Ndeya Makanya or Fokoti Makanya.

cause of no little anxiety in both church and state and warranted close scrutiny by missionaries, statesmen, security forces and scholars alike. Because of their mainly racial orientation the secessionist churches were described as Ethiopian at the turn of the century and the Zulu Congregational Church was referred to as such. For the next two decades the term Ethiopian was reserved for all African independent movements in South Africa which were not favoured by missionaries and by colonists and politicians of the ruling classes, the former because Christian comity was being undermined[2] and the latter because the ascendency and supremacy of the white race were being threatened.[3] It was for this reason that the Bantu Union of the Cape Colony observed that the term Ethiopian was used as a "tar brush" for all African aspirations to make them appear seditious.[4]

Later, as the number of independent church organizations escalated and their characteristics varied, the students of mission science distinguished between Ethiopian churches and other so-called separatist churches.[5] In the middle of

---

[2]Frederick B. Bridgman, Ethiopian movement and other independent factions characterised by a national spirit. *South African deputation papers: presenting some aspects of the work of the two South African missions of the American Board, prepared for the information of the deputation of 1903.* Durban: American Zulu Mission, 1904, p. 24.

[3]Cd. 3247, *Correspondence: re Native disturbances in Natal*: A. J. Hedgeland to J. D. Taylor, Pietermaritzburg, 11 Jun. 1906, p. 33. (British Parliamentary Papers)

[4]*Grievances and memorial with a relative exposition of the Native question submitted to the Right Hon. Prime Minister and to the Union Government of South Africa by members of the delegates of the Bantu Union of the Cape Province, July, 1920*; edited by A. K. Soga. Queenstown: Daily Dispatch Printer, 1920, pp. 32-33.

[5]Amos Burnet, Ethiopianism. *Church missionary review*, Vol. 78, 1922, p. 29. Allen Lea, *The native separatist church movement in South Africa*. Cape Town: Juta, 1926, pp. 18-19. R. G. Norton, Emergence of new religious organizations

the twentieth century, however, the term was employed by B.G.M. Sundkler as a typology of a particular African independent church with distinguishable characteristics.[6] Sundkler's typology has enjoyed wide acceptance among scholars, and over the years the breakaway churches have been a subject of intensive study under a variety of disciplines. The present study is an historical approach and is intended to reconstruct the sequence of events which culminated in the breach between Shibe and the American Zulu Mission and led to the formation of the Zulu Congregational Church. The circumstances of the failure of a reconciliation between Shibe and the Mission, which both parties desired, shall also be recounted. The reconstruction shall be based on the facts gleaned from the available records created contemporaneously with the events.

The Table Mountain mission station, about sixteen kilometres from Pietermaritzburg in Natal, was occupied by the American Zulu Mission in January 1848 when the Rev. Samuel Marsh was posted there on his arrival in the mission field. Marsh, however, left the station in 1850 and the American Zulu Mission employed the Rev. Jakob Ludwig Dohne, originally of the Berlin Missionary Society. Dohne also left the mission station in 1862 to rejoin his parent Society.[7] Soon afterwards the Rev. John Reynolds of the Natal Congregational Union in Pietermaritzburg enquired from the American Zulu Mission if the Union could be allowed to establish a mission of its own at Table

---

in South Africa: a discussion of causes. *Journal of the Royal African society*, Vol. 40, 1940, p. 62.

[6]B.G.M. Sundkler, *Bantu prophets in South Africa*. London: Lutterworth, 1948, p. 53, and, Concepts of Christianity in the African independent churches. *African studies*, Vol. 20, 1963, p. 203.

[7]Lewis Grout, *Zulu-land; or: life among the Zulu kafirs of Natal and Zulu-land*. Philadelphia: Presbyterian Publishing Council, n.d. p. 218.

Mountain.[8] Dohne made a similar enquiry for two years in succession.[9] No firm offer was made to either Reynolds or Dohne. Yet the American Zulu Mission did not place another missionary in that station, but continued to occupy the station and the Mission was given title to 5,623 acres as a mission reserve in 1875.[10] The Mission engaged, first Daniel Mjaleki and later Daniel Ngidi as lay preachers to do pastoral work under the Rev. Stephen C. Pixley's supervision. However, in 1889 the Table Mountain church invited Shibe, a graduate of the Mission's theological college at Adams, to be its preacher.[11]

Shibe was born about 1826 in Franklin, south-west of Durban. He was converted to Christianity after the arrival of the American Zulu Mission in Natal, first worked as a bricklayer and a road maker and later attended school at the Mzumbe Mission Station and was selected to study theology. After seven years he returned to Mzumbe and became a leader of the Christian community in that area. Shibe was one of the few Africans in Natal to be exempted from Native Law and at the beginning of the 1890s he became a member of *Funanamalungelo*, a society of exempted Africans formed in order to petition for civil rights.[12] On being appointed to Table Mountain he worked at the station doing pastoral work as an Approved Preacher and ran a school single-handed for the next six years.

---

[8]ABMC III/3/1, Letters between Missionaries, 1835-1899: J. Reynolds to H. A. Wilder, Pietermaritzburg, 12 Dec., 1862. (Natal Archives)

[9]ABMC III/3/1: J. L. Dohne to Wilder, Wattburg, 24 Sep., 1866; Dohne to Wilder, Umtwalume, 8 Mar. 1867.

[10]Deanne Philippa Collins, *The origins and formation of the Zulu Congregational Church, 1896-1908*. MA (Natal) p. 61.

[11]ABC A/748/USA 81, Letters to Dr Judson Smith: S. Mzoli and E. Zondi to J. Smith, Table Mountain, 10 Nov. 1897. (Transvaal Archives)

[12]Switzer, *op. cit.*, pp. 383-384. Collins, *op. cit.*, p. 65, Note 26 and p. 105.

The Natal Congregational Union had now established an *entente cordial* with the American Zulu Mission, and the Rev. David Russell of the Union visited the Table Mountain mission station for revival meetings.[13] Relations between the Natal Congregational Union and the American Zulu Mission were further strengthened in July 1890 when the Union invited the American missionaries to their annual meetings.[14] Russell continued visiting Table Mountain and the American Zulu Mission instructed the missionary members of *Abayisithupha*, the Committee of Six, consisting of two missionaries and four African members of the Zulu Home Missionary Society, to influence that body to place a man at Table Mountain who could act as an interpreter to Russell during his evangelical work.[15] By this time the Natal Government was pressing the American Zulu Mission to locate a white missionary at the Table Mountain mission reserve.[16] In the middle of 1891 the American missionaries took the initiative and invited the Union to co-operate with them in doing mission work at Table Mountain. Pixley was asked to confer with Russell as to methods and means.[17] Russell agreed and indicated that he wished to have greater control over the mission

---

[13]J. Dexter Taylor, *The American Board Mission in South Africa: a sketch of seventy-five years*. Durban: Singleton, 1911, p. 22.

[14]ABMC III/1/1, Correspondence, 1837-1900: J. Fernie to H. D. Goodenough, Pietermaritzburg, 21 Nov. 1890. (Natal Archives)

[15]ABC A/747/USA 80, General Letters, Annual Reports, Semi-Annual Meetings, Reports - Stations: Semi-Annual Meeting, 31 Dec. 1890. (Transvaal Archives)

[16]H. D. Goodenough, Mission lands: glebes and reserves. *South African deputation papers*, p. 81.

[17]ABC A/747/USA 80: General Letter, 1898.

station.[18] As Russell's plans were in conflict with the Mission's title to the mission reserve further negotiations were necessary. The transfer of control of the mission station, therefore, was delayed until the end of 1895 when the Articles of Agreement were signed by the American Zulu Mission and the Natal Congregational Union, subject to ratification by the Annual Meeting of the American Zulu Mission in July 1896.[19]

The timing was not propitious for the American Zulu Mission and the Natal Congregational Union, the would-be guest mission agency. Discontent among the converts at the Table Mountain station of the American Zulu Mission began mounting from about the middle of 1896 when it became known that the American Zulu Mission wanted to bring that station under the control of a white missionary to the exclusion of an ordained African pastor. The plan to place a white missionary on the station was implemented much to soon after the American Board of Commissioners for Foreign Missions, the mother body of the American Zulu Mission, had joined other Foreign Mission Boards in the United States and Canada at a Conference which decided that it was time the churches in their mission fields became autonomous. The decision of the Conference was conveyed to the mission fields in a Circular Letter dated 14 February 1895. In the Circular Letter the Conference declared that the aim of all Foreign Mission Boards was "to plant a native church, drawing its material support entirely from the native community, which will be ministered by a native pastorate, and will be self-supporting, self-governing and self-propagating." The Conference also expressed itself to be desiring "to see in every land, not an American church or

---

[18]ABC A/749/USA 82, Letters to Dr Judson Smith: Goodenough to Smith, Umvoti, 21 Jul. 1891. (Transvaal Archives)

[19]ABMC I/1/6, Minutes, 1893-1900: Annual Meeting, 11 Jul. 1896. (Natal Archives)

foreign church of any sort but a church native to the soil growing from its roots, and deriving its support from the people among whom it exists."[20]

In a covering letter dispatched with the Circular Letter Dr. Judson Smith, Secretary of the American Board, instructed the American Zulu Mission to make its contents known in all stations of the Mission. He also pointed out that it was a "favorable time to draw out the power of self-support and self-determination in our native churches everywhere."[21] In South Africa the Circular Letter was received in time for the American Zulu Mission's Annual Meeting in July which also resolved that time had come when the missionaries should withdraw from distinctly pastoral work, thus compelling the churches to call and maintain their own pastors. The Letter was translated into Zulu and distributed among the churches and committees connected with the Mission.[22] In the translated version of the Circular Letter the Zulu converts could read:

> "In the opinion of our churches it is found that your churches should be self-governing, should have their pastors whom they should support and should be self-propagating."[23]

---

[20]ABMC II/1/2, Letters and Documents Received from the Board (Boston): *Conference of Representatives of Foreign Mission Board and Societies in the United States and Canada*. New York, 14 Feb. 1895. (Natal Archives)

[21]ABMC II/1/3, Minutes of the Meetings of the American Zulu Mission in South Africa, 1837-1912: Smith to Goodenough, Boston, 21 May 1895. (Natal Archives)

[22]ABMC I/1/6, Minutes of the Minutes of the American Zulu Mission in South Africa, 1837-1912: Annual Meeting, 6 Jul. 1895. (Natal Archives)

[23]ABMC VII/5/1, Native Churches, 1897-1904: *Amabandla aSemerika a Pete abefundisi Ezizweni*, New York, 14 Feb. 1895. (Natal Archives)

The converts took the message of the Circular Letter seriously and acted on it without delay. One of the American missionaries thought this was a step in the right direction and that the Circular Letter was bearing fruits.[24]

When the take-over of the Table Mountain was imminent, the American Zulu Mission erected a chapel at the station. The Table Mountain people, on their side, co-operated in building a house for the use of the missionary.[25] The Natal Congregational Union appointed the Rev. John G. Pugh to be its missionary at Table Mountain, and he and his family were already at the station by the middle of February, 1896, though the transfer was to be effective only on 22nd April.[26] The house was not ready, however, when Pugh arrived and in his first report to the American Zulu Mission Pugh expressed his regret that so much of his time had been taken in putting up the house.[27]

Late the previous year Shibe had felt unwell and had indicated his desire to take a rest from the mission station and to receive medical treatment. To that end he had applied for leave of absence from duty from the church at Table Mountain, the American missionaries and from *Abayisithupha*.[28] He was given four months leave with effect from 22nd April, 1896, and was advised to obtain

---

[24]ABC A/749/USA 82, Letters to Dr Judson Smith: C. J. Dorward to Smith, Umzumbe, 18 Jul. 1895 (Transvaal Archives)

[25]ABMC IV/1/2, Committee Reports of the AZM, 1896-1899: Committee on Table Mountain, 1896. (Natal Archives)

[26]ABMC III/7/1, Letters: Rev. C. W. Kilbon, 1896-1899: Kilbon to J. G. Pugh, Adams, 15 Feb. 1896. (Natal Archives)

[27]ABC A/747/USA 80: Annual Report on Table Mountain, 1896.

[28]ABMC VII/5/1, Diverse Matters: Native Churches, 1897-1904: S. B. Shibe to Kilbon, Lovedale, 27 Apr. 1896. (Natal Archives)

permission from Pugh as well.[29] When Pugh took residence at Table Mountain, he was informed by the American missionaries about Shibe's leave, and Pugh had intimated his willingness to allow Shibe to depart.[30] Shibe went to the Lovedale Hospital in the Eastern Cape where he was attended to by a certain Dr. Kelbe.[31]

Even before the American Zulu Mission could ratify the Articles of Agreement for the transfer of the Table Mountain station to the Natal Congregational Union, Pugh had indicated his disapproval of Shibe and protested his inability to work with him.[32] He also requested the American Zulu Mission to transfer Shibe to another station on his return from Lovedale, and that he, Pugh, should be allowed to choose a new man in Shibe's place.[33] He gave no reasons for his requests. Pugh's requests took the Mission by surprise and were referred to the Committee on Table Mountain and that Committee recommended to the Annual Meeting that the parties concerned should be consulted before any action, and that if the church at Table Mountain wished to call Shibe or any Approved Preacher and assume the entire support such a man "be ordained pastor of that church."[34]

To ensure widest consultations the Annual Meeting, in turn, referred the matter to the Native Agency Committee, which included two African pastors who were members of *Abayisithupha*. That Committee granted Pugh's request and

---

[29]ABMC III/7/1: Kilbon to Shibe, Adams, 15 Feb. 1896.

[30]*Ibid.*: Kilbon to Pugh, Adams, 15 Feb. 1896.

[31]ABMC VII/5/1: Shibe to Kilbon, Lovedale, 10 Jun. 1896.

[32]ABMC IV/1/2: Native Agency Committee, 1896.

[33]*Ibid.*: Committee on Table Mountain, 1896.

[34]*Ibid.*

appointed Shibe to Noodsberg, another of the Mission's stations.[35] Shibe was advised about the decision while he was still at Lovedale and he indicated his willingness to comply.[36] But on hearing the decision to transfer Shibe, the church members at Table Mountain accused the Mission of having sold them to the Natal Congregational Union as they had not been consulted.[37] They wanted to retain Shibe as their preacher and sent Samuel Mzoli and Elijah Zondi to urge the Mission to allow Shibe to remain at Table Mountain. They indicated that the church was willing to assume Shibe's entire support, and, on the strength of the Circular Letter of 14 February 1895, they also requested to have Shibe ordained as their pastor.[38] Mzoli and Zondi further communicated the events at Table Mountain to Shibe, and the latter, while still at Lovedale, now indicated his willingness to accept a call from the church to which he felt bound.[39] The African pastors, including the two who were members of *Abayisithupha*, also espoused Shibe's and the Table Mountain church's cause.[40] Towards the end of December 1986 Shibe returned to Natal by way of Durban. On the 23rd of December, while still in Durban, he was met by Pixley with a message from Pugh that he, Shibe, should no longer go to Table Mountain. Protesting that the ban amounted to his being separated from his wife and family Shibe defied the order

---

[35]ABC A/747/USA 80: F. Bunker's Report on a Visit to Table Mountain, 1896.

[36]ABC A/751/USA 84, Letters to Dr Judson Smith: S. C. Pixley to Smith, Inanda, 25 Jan. 1898. (Transvaal Archives)

[37]ABC A/747/USA 80: Mzoli and Zondi, Table Mountain, 10 Nov. 1897.

[38]*Ibid.*

[39]ABMC VII/5/1: Shibe to Kilbon, Table Mountain, 22 Sep. 1898.

[40]ABC A/747/USA 80: General Letter, 1898.

and went to Table Mountain, only to find that Pugh had destroyed some of his property.[41]

At the beginning of 1897 the crisis at Table Mountain merged with another crisis in Johannesburg, where the American Zulu Mission had extended itself in 1893 in order to minister to the spiritual needs of its converts who had gone there for employment. Control of the church , its property and its funds, as well as placing of an African pastor over the church were the main causes of strife in Johannesburg.[42] The matter had been discussed by the Annual Meeting of the Mission in July 1896. The Meeting backed the American missionary in Johannesburg, the Rev. Herbert G. Goodenough, in limiting the church members' participation in the disbursement of church funds and the missionary's restriction of Fokoti Tyoyo, whom the church wanted to have ordained as a pastor, to an out-station at Elandsfontein.[43] These strictures incensed the church members who informed the African pastors in Natal about their plight. The pastors took the Johannesburg matter up. The church in Johannesburg felt strengthened and again asked Goodenough to instal Tyoyo as their pastor, but the missionary refused.[44]

When the Mission's Semi-annual Meeting assembled in January 1897, the Natal Congregational Union had been urging the Mission to remove Shibe from Table Mountain, and the churches at Table Mountain and in Johannesburg had

---

[41]ABMC VII/5/1: Shibe to Kilbon, Table Mountain, 25 Jul. 1898.

[42]A. W. Z. Kuzwayo, *A history of Ethiopianism in South Africa, with particular reference to the American Zulu Mission from 1835 to 1908*. MA (Unisa) 1979, pp. 103-131.

[43]ABC A/749/USA 82: Goodenough to Smith, Johannesburg, 15 Apr. 1897.

[44]ABMC (Unclassified): Meeting with Native Pastors, 22 Jun. 1897.

been pressing for Shibe's and Tyoyo's ordination as pastors.[45] In response the American Zulu Mission merely conveyed the Table Mountain's request to the Natal Congregational Union, but resolved to inform "certain natives in Johannesburg" that Tyoyo "should work where in the opinion of the missionary they were needed most."[46] The response drove the pastors to convene a meeting with preachers and delegates from the churches in the Mission without the customary prior permission of the missionaries. The meeting was held at Adams on 17 March 1897. Relying upon the Circular Letter of 14th February 1895, they sent their circular letter to the churches criticising the Mission's strictures on Shibe and Tyoyo and commending the Secretary of the American Board for the Circular Letter of 1895.[47] Shibe and Tyoyo were encouraged by the support from churches and displayed their opposition to the missionaries placed over them. Shibe would not go to Noodsberg and Tyoyo told Goodenough in the face that he was leaving Elandsfontein, and both ran rival services in opposition to those held by the missionaries.[48]

Faced with the recalcitrant churches at Table Mountain and in Johannesburg the American missionaries convened their Annual Meeting for 1897 as early as 16 June, and invited the only five pastors of the Mission for discussions in committees. The pastors clearly indicated their support for Shibe, backed the church in calling him as their pastor in the spirit of the Circular Letter of 14 February 1895 and rejected the excuse that Pugh's presence at Table

---

[45]ABMC I/1/6: Annual Meeting, 7 Jul. 1896.

[46]*Ibid.*: Semi-Annual Meeting, 28 Jan. 1897.

[47]ABC A/749/USA 82: Goodenough to Smith, Johannesburg, 15 and 18 Apr. 1897.

[48]ABMC II/1/1: W. Searle to Goodenough, Pietermaritzburg, 1 Apr. 1897. ABC A/749/USA 82: Goodenough to Smith, Johannesburg, 18 Apr. 1897.

Mountain was dictated by the Natal Government's demand for a white missionary presence on the reserve.[49] The pastors also backed the Johannesburg church in demanding Goodenough's removal from Johannesburg and the return of monies paid by church members for buildings.[50]

When the missionaries met separately in the Annual Meeting they disregarded the Johannesburg church's complaints on church property and financial matters, but decided to remove Tyoyo's name form the list of Approved Preachers, so that he could no longer be considered for ordination, because he had disobeyed the missionary.[51] On Table Mountain matters the missionaries paid attention to the Natal Congregational Union's insistence that Shibe be evicted from the mission station and recommended that the Union should lay charges against Shibe to be investigated by an impartial committee of two pastors, two American missionaries and two representatives of the Union.[52] The matter was investigated in September in Durban, but the pastors were not part of the team and Shibe was not invited to defend himself. The charges did not reveal why Pugh disapproved of Shibe and referred to nothing that took place after Pugh's arrival at Table Mountain, except that Shibe had gone to Lovedale for a lengthy stay when he had been given four months leave. The committee's recommendations amounted to a vote of confidence in Pugh and to a demand for Shibe's subservience to Pugh on the pain of being evicted from Table

---

[49]ABMC (Unclassified): Meeting with Native Pastors, 23 Jun. 1897.

[50]ABMC IV/1/2: Committee on Johannesburg Work to Makubalo and others, Adams, 7 Jul. 1897.

[51]ABMC I/1/6: Annual Meeting, 7 Jul. 1897.

[52]*Ibid.*: Annual Meeting, 8 Jul. 1897.

Mountain.[53] On receiving the verdict Shibe requested a hearing with the missionaries, but the latter, who were now embarrassed by the conduct and the results of the investigations,[54] did not respond.[55] The church at Table Mountain reacted with hostility towards Pugh and demanded that he should vacate the house at the mission station.[56] Reacting to an inquiry from the Secretary of the American Board in Boston on complaints lodged by Mzoli and Zondi,[57] Pixley disputed nothing but opined that Pugh was prejudiced against Shibe.[58]

On getting no satisfaction from the missionaries Shibe approached *Abayisithupha* requesting a meeting. The two African members of *Abayisithupha* then took it upon themselves to convene another meeting of African pastors, preachers and delegates from churches, despite the ban on such meetings by the American missionaries,[59] and the group met at Thafamasi towards the end of 1897.[60] The Thafamasi meeting recommended that Shibe should remain at Table Mountain, avoiding any quarrel with Pugh, and that the church should raise a purse for Shibe's support. The meeting also advised the Johannesburg

---

[53]ABMC VII/5/1: Report of the Table Mountain Matters Dealt with in Congregational Union Meeting, Durban, 15-16 Sep. 1897.

[54]*Ibid.*: Kilbon to Fernie, Adams, 27 Sep. 1897.

[55]*Ibid.*: Shibe to Kilbon, Durban, 2 Aug. 1898.

[56]ABMC (Unclassified): *Ibandla* to Pugh, Table Mountain, 29 Oct. 1897.

[57]ABC A/747/USA 80: Mzoli and Zondi to Smith, Table Mountain, 10 Nov. 1897.

[58]ABC A/751/USA 84: Pixley to Smith, Inanda, 25 Jan. 1898.

[59]ABMC I/1/6: Annual Meeting, 8 Jul. 1897.

[60]ABC A/747/USA 80: General Letter, 1898.

group to form themselves into a church and to avoid quarrelling with Goodenough.[61] With the backing of pastors, preachers and delegates of churches assured for their cause, Shibe went to Johannesburg in December to confer with Tyoyo in connection with the formation of a church there and to arrange for their ordination as pastors.[62]

On hearing about the decision of the Thafamasi meeting and the purpose of Shibe's departure for Johannesburg, the American Zulu Mission instructed the Rev. C.W. Wilcox to write a letter to the pastors informing them that what was contemplated would constitute a breach with the Mission, and to request them to discuss these matters with the American Zulu Mission the following January.[63] The meeting did take place and resulted in the appointment of the Johannesburg Emergency Committee, consisting of missionaries and pastors, to meet and work for reconciliation. However, the committee did not arrive in Johannesburg, where Shibe and Tyoyo were expecting them, until it was too late to achieve its purpose.[64] Having been denied the influence of the missionaries, Shibe and Tyoyo now took legal counsel with H. Cohen,[65] a Jewish lawyer, much to the annoyance of the Christian missionaries, who advised them on the legal aspects of the formation of a church and the ordination of pastors, all for a fee of five pounds.[66] On the 14th February 1898 they convened a meeting consisting of ninety members of the Mission converts and adopted a document

---

[61]ABMC IV/1/2: Native Annual Meeting Committee Report, Feb. 1898.

[62]ABMC VII/5/1: Pugh to J. L. Hulett, Table Mountain, 10 May 1898.

[63]ABMC III/3/1: C. W. Wilcox to Native Pastors. [s.l.], 1897.

[64]ABMC IV/1/2: Johannesburg Emergency Committee Report, 1898.

[65]ABMC VII/5/1: Shibe to Kilbon, Table Mountain, 22 Aug. 1898.

[66]ABC A/749/USA 82: Goodenough to Smith, Johannesburg, 3 May 1898.

which was later referred to as the "First Charter of Independence" and called it the "Covenant of the Zulu Congregational Church." It read:

> "We, the undersigned, being the members and founders of the above named church, have agreed by resolution to separate ourselves from the American Congregational Missionary Society(sic). Let it be known, therefore, that as from the 14 February, 1898, we have separated ourselves altogether from the American Congregational Missionary Society(sic). We have founded our own church which is the Zulu Congregational Church."[67]

All ninety members present appended their signatures to the document. The new church took early steps to have its leaders ordained as pastors on the 27th February 1898,[68] apparently with no regard for apostolic succession.

When the American missionaries heard about the formation of a new church and Shibe's and Tyoyo's ordination they called a Special Meeting for 28th and 29th April, 1898, to devise ways of isolating the new organization. A message was sent to loyal churches showing that the principles of congregational usage had been violated and warned them not to enter into any alliance with the Zulu Congregational Church.[69] By the middle of year it was evident that the missionaries had been shaken by the recent events. The Annual Meeting of the Mission approved a message to the churches in which the American Board's mission policy of self-supporting, self-governing and self-propagating churches was reiterated and regretted that some members had broken away prematurely.

---

[67]*Covenant, constitution and deed of trust of the Zulu Congregational Church*, 24 Mar. 1903. (MS. Zulu Congregational Church, Sobantu Village.)

[68]ABC A/749/USA 82: Goodenough to Smith, Johannesburg, 9 May 1898.

[69]ABMC I/1/6: Special Meeting, 28 Apr. 1898.

The General Letter of the year stated that the old relation between the missionary and his church must yield to sympathetic influence.[70]

Pugh, however, was now even more determined to oust Shibe from the Table Mountain mission reserve as he found his stay untenable.[71] He wrote to the Secretary for Native Affairs requesting him to have Shibe removed from Table Mountain,[72] and informed Kilbon that Shibe was a trespasser at Table Mountain and that the Secretary for Native Affairs felt that the matter had resolved itself into a political question as Shibe was influencing the minds of the Africans.[73] Hoping that Shibe could still be persuaded to rejoin the Mission, the 1898 Annual Meeting of the American Zulu Mission instructed Kilbon to summon Shibe to sign a document renouncing his ordination and agreeing to submit to the Mission. Failing that the Mission would consider evicting him from the reserve.[74] It was while the Annual Meeting was in session that word reached the Mission that Tyoyo had died in Johannesburg.[75]

Kilbon did invite Shibe to meet him, first in Pietermaritzburg on 6 July and later at Manzimtoti on 11 July, 1898.[76] When Shibe failed to keep both appointments the Annual Meeting decided to drop Shibe's name from the list of

---

[70]ABC A/747/USA 80: General Letter, 1898.

[71]ABMC VII/5/1: Pugh to Kilbon, Table Mountain, 5 May 1898.

[72]*Ibid.*: Pugh to J. L. Hulett, Table Mountain, 16 May 1898.

[73]*Ibid.*: Pugh to Kilbon, Table Mountain, 16 May 1898.

[74]ABC A/747/USA 80: Annual Meeting, 1898.

[75]ABMC I/1/6: Annual Meeting, 28 Jun. 1898.

[76]ABMC VII/5/1: Shibe to Kilbon, Table Mountain, 11 Jul. 1898.

Approved Preachers.[77] Judging by the letter Shibe wrote on 22 August, Shibe misunderstood the missionary's intentions when Kilbon invited him to a meeting. In that letter Shibe apologised profusely:

> "Even though my father blames me," he wrote in part, "I know that I hold nothing against the missionaries... I did not conspire to separate myself from you. Had I conspired to separate myself from you I would have joined other denominations.
>
> "When I hear that you wish the work to continue in harmony and in seriousness, [I have to state that] I bear no ill-feeling against you. I am the product of your hands. Even though you blame me, you are within your rights because there is no child who is not under the censure of his father."[78]

His belated apologies were to no avail in that the American Zulu Mission had already instituted legal action to have Shibe evicted from Table Mountain. Shibe lost the case before the Magistrate's Court, but on appeal to the Supreme Court Shibe won.[79] This was a severe set-back to the missionaries and to Pugh, who lost face.

From the middle of 1899 matters took a new turn in the American Zulu Mission field. At the June Annual Meeting of the pastors and missionaries, which had since been revived, the Table Mountain branch of the Zulu Congregational Church sent a delegation which was prevented from participating

---

[77]ABC A/747/USA 80: Annual Meeting, 11 Jul. 1898.

[78]ABMC VII/5/1: Shibe to Kilbon, Table Mountain, 22 Aug. 1898. [A free translation from the Zulu original]

[79]Shibo(sic) v. American Board of Missions (sic). *Natal law reports*, Vol. 19, 1898, pp. 175-176.

in the meeting only because it could not produce a letter of invitation.[80] This led to a repetition of a suggestion that a united and earnest effort for reconciliation between the Zulu Congregational Church and the American Zulu Mission be made. The meeting recommended that a committee of ten, consisting of five members from the Zulu Congregational Church, three from the Zulu Home Missionary Society and two white missionaries, be constituted to work for reconciliation.[81] The American Zulu Mission sent an invitation to the Zulu Congregational Church for a meeting to be held in Durban on 22 August 1899. Shibe and his party went to Durban on the appointed day, only to find that the meeting had been cancelled.[82]

The Joint Meeting of the Representatives of the American Zulu Mission, the Zulu Home Missionary Society and the Zulu Congregational Church began in Durban on the 14th November 1899. The Zulu Congregational Church was represented by Shibe and Nyuswa, who had since taken the leadership of the Church in Johannesburg after Fokoti's demise, and three other members to match the five adherents of the American Zulu Mission churches. A sub-committee of four, on which both Shibe and Nyuswa sat, was formed to draw up points which they judged to be standing in the way of union and co-operation.[83] The adjourned Joint Meeting re-assembled on the 30th of January 1900 and the sub-committee presented its report which purported to bring out the "exact grievances that must be righted." But during the early stages of discussions Nyuswa informed the meeting that the Zulu Congregational Church was willing

[80]ABC A/747/USA 80: General Letter, 1898.

[81]ABMC IV/1/2: Committee on Native Annual Meeting Report, 1899.

[82]ABMC VII/5/1: Shibe to Kilbon, Table Mountain, 1 Sep. 1899.

[83]ABMC IV/1/4: Committee on Reconciliation of the Zulu Congregational Church Report, Jun. 1900.

to let the past be forgotten if the missionaries would accept a scheme which he proposed. The scheme purported to shift the emphasis from points of differences to a practical scheme setting out conditions for re-union. The proposals were that all the churches under the American Zulu Mission had to govern themselves with their pastors, while the missionaries only helped where necessary. The Zulu Home Missionary Society was to be trustee of the church lands not in the mission reserves, and the name Zulu Congregational Church was to be retained by all the churches which have sprung up by themselves. There was to be some liaison between the American Zulu Mission and the Zulu Congregational Church.[84]

By presenting what the American missionaries regarded as a "skilful plan"[85] Nyuswa emerged as a leader of the Zulu Congregational Church and was accepted by the missionaries as its spokesman. He and the members of the Johannesburg group accepted the fact and did not take the trouble to obtain the views of the Table Mountain branch. When the Americans presented a counter-scheme which modified Nyuswa's during the resumed discussions later in June, Nyuswa accepted the counter-scheme on behalf of the Zulu Congregational Church and closed the discussions. It was evident, however, that the Table Mountain people were no longer so well disposed to the Johannesburg party.[86]

All three parties of the Joint Meeting undertook to present the proposals to their parent bodies for ratification. The Annual Meeting of the American Zulu Mission accepted the proposal in June,[87] while the Zulu Home Missionary Society accepted them in their Native Annual Meeting, at which the Zulu

---

[84]ABMC VII/5/1: Scheme Presented by Rev. Sunguza Nyuswa, 30 Jan. 1900.

[85]ABMC I/1/6: Semi-Annual Meeting, 7-8 Feb. 1900.

[86]ABMC IV/1/4: Committee on Reconciliation of the Zulu Congregational Church Report, Jun. 1900.

[87]ABMC I/1/6: Semi-Annual Meeting, 7-8 Feb. 1900.

Congregational Church and the representatives of the American Zulu Mission were present in July. The Zulu Congregational Church had not reached unanimity as Shibe still had reservations. However, Banana Makanya of the Johannesburg church, read the official statement of the Zulu Congregational Church:

> "We, the majority of the Zulu Congregational Church and including the leader thereof, consent that the brotherly work which you propose, and which is contained in the scheme of *Mfundisi* Nyuswa, proceed. But we wish you should know that there are a few of the people of the Zulu Congregational Church who have a wish - and *Mfundisi* Shibe is one of them - though even he says that what the majority decided upon he will agree to. Most of them concur in that which we have adopted in the meeting which we have been holding."[88]

The two branches of the Zulu Congregational Church were advised each to call a council for the purpose of forming a church to be recognised by the American Zulu Mission as sister churches to the rest of the Mission churches. However, Shibe did not accept the invitation to attend the meeting for reconciliation when the Johannesburg church was being recognised, nor did he call a council for the recognition of his branch. To Kilbon he indicated that he had decided not to reunite with the other churches and he wished to be left alone. But to Pugh he gave out that in view of the work he had begun in several places he could not commit himself about the union until he had consulted his followers at the next Annual Meeting of his church. In the event Shibe never reunited himself and his church with the Natal churches nor with the American Zulu Mission. In order to differentiate between Shibe's Zulu Congregational

---

[88]ABMC IV/1/4: Committee on Reconciliation of the Zulu Congregational Church Report, Jul. 1900.

Church and their new organization, the Natal and the Johannesburg churches adopted the name African Congregational Churches.[89]

The *prima facie* case emerging in the narrative given above is that the primary cause of the breach between the members of the churches of the American Zulu Mission, on the one hand, albeit that the main role players were at Table Mountain and in Johannesburg, and the American Zulu Mission, on the other, were the accumulated effects of the reversal of the policies allowing for a greater measure of self-government among the churches without the intervention of the missionaries. Closely related to control by missionaries was the reluctance on the part of the Mission to ordain more pastors in the mission field.

Already in 1861 the Native Home Missionary Society, consisting exclusively of African converts, had been formed in the spirit of the three-selves formula for the planting of self-supporting, self-governing and self-propagating churches in the mission field.[90] The body collected funds for the support of helpers or preachers or pastors in mission field and held annual meetings on a variety of issues. To the funds of the Society the American Board contributed grants-in-aid to supplement salaries of helpers and preachers.[91] In 1885 the American Zulu Mission came up with a plan of *Abayisithupha* which included two missionaries to be an executive committee of the Native Home Missionary Society so that the Mission should have a voice in the management of the Society "so long as they furnished any part of its funds, and no longer."[92] However, in 1894 the American

---

[89]*Ibid.*

[90]Taylor, *op. cit.*, p. 74.

[91]ABMC II/1/2: R. Anderson to AZM, Boston, 27 Nov. 1861.

[92]ABMC III/6/2: E. Robbins to Smith, Adams, 1 Feb. 1886.

Board ceased to contribute to the treasury of the Society,[93] yet the two missionaries remained on *Abayisithupha* and exercised some influence in the government of the churches. Thus when Pugh requested the removal of Shibe from Table Mountain in 1896, that committee granted the request without sensing the views of the church in question. Indeed the American Zulu Mission involved the Congregational Union at Table Mountain without taking the church into confidence.

If missionary influence in the government of churches was restrictive, the Mission's failure to ordain pastors and to place them over the churches was an aggravating circumstance. When the controversies began there were nineteen[94] churches that had been founded in the American Zulu Mission field. While there were eight ordained African pastors in 1883,[95] there were only five pastors in active ministry in 1897. Others had either died, simply abandoned the church or had been defrocked and not replaced. Even though the American Board was in no position to increase its staff in the field the American Zulu Mission resisted all pressure for the ordination of pastors over the churches. At Table Mountain they virtually handed the mission station to another agency which appointed a white missionary to exercise control over the church and arranged to transfer

---

[93]C. W. Kilbon, *Abaisitupa*; or, the Committee of Six in Home and Foreign Mission work. *South African deputation papers*, p. 17.

[94]Taylor, *op. cit.*, p. 24, reports 15 pastors in 1885. Four more had been ordained in 1897. Cf. A. C. Maseko, UMafukuzela, *Ilanga lase Natal*, 13 Feb. 1946. ABC A/747/USA 80: General Letter, 22 Dec. 1891. Arthur Frijof Christopherson, *Adventuring with God: the story of the American Board Mission in South Africa*; edited by Richard Sales. Durban: Robinson, 1967, p. 91. ABMC III/1/1: S. Nyawose to Goodenough, Umtwalume, 25 Jan. 1895.

[95]Edwin W. Smith, *Life and times of Daniel Lindley, 1801-1890: missionary of the Zulus, pastor of the Voortrekkers, ubebe omhlophe*. London: Epworth Press, 1929, p. 400-401.

Shibe, who was being called by the church, to another mission station. In Johannesburg the American Zulu Mission resisted the church's request for a pastor, but appointed a local white lady as a kindergarten teacher for the benefit, mainly, of illegitimate children of mixed parents who were not even communicants of the church.[96]

The American Zulu Mission had gone back on the American Board's fundamental mission policy of planting self-supporting, self-governing and self-propagating churches under the control of a native pastorate. As far back as 1860 the Secretary of the American Board, Dr. Rufus Anderson (1832-1864), had urged the American Zulu Mission to place native pastors in all churches, and that the responsibility of government should be thrown upon the churches.[97] In 1867 his successor, Dr. Nathaniel G. Clark (1865-1884), had reminded the missionaries that their relation with the churches was only temporary, and that they were to leave the churches in the hands of the pastors over whom the missionaries would have no authority.[98] The same message was repeated by Dr. Judson Smith in the Circular Letter of 14th February 1895. Despite repeated instructions from a succession of the American Board's secretaries the missionaries had not only failed to ordain more pastors, but in the General Letter of 1897 they went as far as to reserve to themselves the right to determine when to hand over the control of the churches to the pastors. They stated that:

> "... We feel that at the same time that it should be fully explained what are our respective relations to the Board and themselves, also that fields long occupied by the Board, must, from the nature of the case, be under the general supervision and oversight of

---

[96]ABMC I/1/6: Semi-Annual Meeting, Feb. 1895.

[97]ABMC II/1/2: Anderson to AZM, Boston, 20 Apr. 1860.

[98]Smith, *op. cit.*, p. 397.

> ourselves as agents of the Board till such time as *we*, and not they, shall deem it safe in the interests of Christ's kingdom to turn the work into their hands.[99]

For the long determined policies of the American Board to be reversed unilaterally by the American Zulu Mission at a time when the American Board had even terminated its grants-in-aid to the churches, and after the distribution of the Circular Letter which reiterated the American Board's commitment to planting self-supporting, self-governing and self-propagating churches in the mission fields it is understandable that even the pastors, preachers and delegates supported the dissident members at Table Mountain and Johannesburg at the Adams and Thafamasi meetings. It is understandable also that Shibe, who was at the centre of the controversies and the only member who suffered the worst treatment by the missionaries, should have cited certain aspects of the policies of the American Board to justify the formation of the Zulu Congregational Church separately from the American Zulu Mission instead of his personal deprivations. He told the Natal Lands Commission of 1900-1902 that:

> "The cause of the dispute was a letter which came from the American Board telling the White missionaries that they were to appoint the Native Missionaries(sic) to work in their congregations... The impression that the letter gave us was that we were no longer to be under their control because it said, 'You must get down off our backs... We have been carrying you long enough... ' The separation did not happen because we refused to be controlled by the White missionaries, but because the people refused to allow the White missionaries to drive them away from Table Mountain."[100]

---

[99]ABMC V/1/1: General Letter, 1897.

[100]Natal Lands Commission, 1900-1902. *Minutes of evidence*. Pietermaritzburg: Government Printer, 1902, p. 9.

If, therefore, consideration is given to the full implications of the official policies of the American Board, it should be accepted that the decisions of the Adams and Thafamasi meetings of the pastors, preachers and delegates were in the spirit of carefully developed mission policies which the American Zulu Mission were expected to implement. By the same token also, it should be accepted that Shibe and Tyoyo intended to give full expression to policies known to all converts when they sought to form separate churches over which they were to be ordained as pastors. Indeed the document, "Basis of Reconciliation between the Churches of the American Zulu Mission and the Zulu Congregational Church" approved at the meeting of the 11th September 1900 and accepted even by the American missionaries as *Magna Carta* for all African churches under the American Zulu Mission,[101] was the work of a joint meeting of the Zulu Congregational Church, the loyal churches and the American missionaries, the latter having since shown a willingness to heed the implications of the policies of their mother body. When the document was adopted, both the loyal churches and the missionaries had accepted the ultimate necessity of allowing the congregations to be self-supporting, self-governing and self-propagating in the spirit of a long standing mission policy developed in the nineteenth century. From the point of view of mission policy the formation of churches under the leadership of the American Zulu Mission converts at the turn of the nineteenth century must be accepted to have been an adumbration of the 'three-selves formula' of the mission agencies, albeit that certain missionaries thought their mission converts acted prematurely.

As for Shibe's failure to return to the American Zulu Mission fold, it is fair to conclude that it stemmed from his hurt feelings, mainly due to Pugh's

---

[101]ABC A/752/USA 87, South African Mission: Zulu, Rhodesia, Beira: Annual Letter, 1900-1901.

prejudice against him, the destruction of his property by Pugh, the litigations he underwent at the hands of the Natal Congregational Union and the American Zulu Mission. The last straw was Nyuswa's shabby treatment of him which the American missionaries also failed to heal. Not only did Nyuswa usurp the leadership of the Zulu Congregational Church and made himself its spokesman, but he actually overruled the tabling of a report of a sub-committee, of which Shibe was a member, without consultation and poured cold water over Shibe's case for the amelioration of the "exact grievances" of the church at Table Mountain. The American missionaries, on their side, accepted the leadership of a man whose involvement in the Zulu Congregational Church had been recent, if only to achieve reconciliation. Even Kilbon, one missionary who was sensitive to the feelings of the African churchmen, had not been impressed with Nyuswa's performance since the Native Annual Meeting in July 1900, and when Shibe did not accept the invitation to the reconciliation of his congregation, Kilbon opined that "(t)here is, therefore, no less hope of future union between the Shibe party and the Natal Churches than there was two years ago between the Johannesburg party and these churches."[102]

If Kilbon had hopes of reconciliation, subsequent developments ensured that Shibe could not be enthused towards the American Zulu Mission. Within months of the signing of the document of reconciliation reports were circulating within the Mission circles that the Zulu Congregational Church had abandoned the high moral standards set in the Msunduze Rules and that Shibe, once a total abstainer, was now carrying a large snuff box.[103] Pressure to remove him and his church from Table Mountain did not abate either. A fraternity of

[102]ABMC IV/1/4: Committee on Reconciliation of the Zulu Congregational Church Report, Jan. 1901.

[103]Bridgman, *op. cit., p. 23-26.*

missionaries painted him and his followers as villains who had undermined Christian comity and accused him of proselytizing and of reaping where others had sowed[104] and, therefore, denied him Christian fellowship.

Despite the failure of a *rapprochement* between the Zulu Congregational Church and the American Zulu Mission, what hostilities that existed between the two bodies in 1898 had abated considerably in September 1900 when the document which formed the basis of reconciliation was drawn. The Zulu Congregational Church, therefore, stepped out, to use the Rev. G.B.A. Gerdener's phrase, as a "legitimate daughter church"[105] of the American Zulu Mission.

Shibe's Zulu Congregational Church has survived as an independent church for close to a century now, much longer than its sister churches, the African Congregational Churches. The latter succumbed to the onslaughts of the Natal Government in 1907 when they subordinated themselves to the American Zulu Mission, taking the name Congregational Churches of the American Board. They even applied for membership of the Congregational Union of South Africa in order that "they be identified with the English colonial Congregational Churches."[106] This was just a truce for some churchmen. In 1917 Gardener Mvuyana, erstwhile an Approved Preacher under Goodenough in Johannesburg, broke away from the American Board and revived the old name, calling his

---

[104]ABC A/754/USA 87: Annual Letter, 1900-1901.

[105]G. B. A. Gerdener, *Recent developments in the South African mission field*. Cape Town: N. G. Kerk Uitgewers, 1958, p. 155.

[106]South African missions, Zulu branch: an eventful meeting. *Missionary herald*, Vol. 103, 1907, pp. 533-535.

church the African Congregational Church.[107] But not only did Shibe's Zulu Congregational Church thrive at Table Mountain and in other American Zulu Mission stations, it even spread to the Transvaal within a decade. If the formation of the Zulu Congregational Church, like *Isonto Lohlanga* in 1885[108] and the African Congregational Church in 1917, undermined Christian comity and if ecumenism was complicated by its continued existence the Church was, nonetheless, instrumental in spreading the Christian gospel to the areas which the American Zulu Mission had not reached. It became, as the Circular Letter of 14th February 1895 pleaded, a church native to the soil growing from its roots and deriving its support from the people among whom it exists.

[107]J. Dexter Taylor, *One hundred years of the American Board in South Africa, 1835-1935*. [Private Print], p. 56.

[108]For the formation of *Isonto Lohlanga*, see Kuzwayo, *op. cit.* pp. 75-81.

CHAPTER 6
A REFORMED AFRICAN INDIGENOUS CHURCH?

by

R.J.R. van der Spuy

Is it possible that an African indigenous church can have a pronounced Reformed character?

In the opinion of Frank Ntsuntsha, as quoted by Dr Johannie Cronje, such a possibility is rather remote. Ntsuntsha, himself a member of an African indigenous church, maintains that African Independent/Indigenous Churches (AIC) proclaim a diluted and truncated Bible message. He came to this view only after his conversion. At present he is involved with the Christ the Rock Indigenous Church Association (CRICA) where leaders of the indigenous churches are trained to enable them to understand and convey the Bible message. He is outspoken about his perception that many of the indigenous churches possess zeal, but without knowledge. In his opinion there are about 3000 adherents of Zionist churches who do hear the "pure" gospel. This view of his is based on an event at a seminar during which about 100 bishops and leaders of indigenous churches committed themselves to Christ as Lord (Cronje 1990:22-23)

Statements like these contributed to the perception that there is virtually no indigenous African church which is really and truly church in the biblical sense. Kruger in his thorough study on the Zionist Christian Church (ZCC) concluded that this movement does not display the attributes of a true church (Kruger 1983:12).

Classification of African indigenous churches

In the light of these views it is important to try to classify the African indigenous churches and religious movements. We cannot consider them all as churches. On the other hand, we cannot classify all of them as being traditional movements with some Christian content.

Turner, a renowned scholar in this field, attempted to provide a classification system by which the African indigenous churches and movements can be evaluated. He placed all the modern African religious movement on a continuum. All types of religious movements appear on this line to assist in evaluating a specific movement (Turner 1968:5). Gilliland (1986:266-272) and others also produced similar classification systems, which are mostly extensions or modifications of Turner's one.

The term "modern African religious movements", according to Turner, embraces the full spectrum of religious movements. This spectrum covers the whole range from traditional religious movements, on the one hand, to the real Christian churches, on the other. He uses three categories, namely Neo-Pagan, Hebraistic and Christian movements. The latter is again subdivided (1968:1-34). For this article our interest lies in the third category.

Turner divides the Christian churches or movements in Africa into two sub-categories: the large established churches with Western origin or connections and the African Independent/Indigenous Churches.

The first sub-category, the established, historical or mainline churches as they are known, are usually of Western origin and connected with leaders from the white population groups. One can call these Churches with Western Connections (CWC). It can be argued that this Western connection constituted the major difference between the CWC and the AIC.

Although the CWC in Africa is meant for the people of Africa, they often cling to the transmitted Western liturgies, church polity and ecclesiastical customs and traditions to such an extent that there is often virtually no indigenization found in them. These include all the large established churches such as the Roman Catholic Church, the Anglican Church, the Methodist Church, the Lutheran Church and the churches which grew out the three Afrikaans churches of Reformed persuasion.

The second sub-category, the AIC, comprises the African churches and religious movements which have no Western ties or connections, but which are indigenous and operate independently from the Western-related churches. Although the terms "churches" and "movements" are often heard when reference is made to the African indigenous religious groups, these groups see themselves as "churches".

The point to be made is that this second sub-category covers churches or religious movements which are independent from Western leadership, ties or even Western influence. Africans formed their own churches to address and serve the African people. The liturgies, customs, usages and agendas are tailored to suit the needs and the character of the African people. For this reason these churches are conspicuously different from the churches with Western background and connections.

Sundkler, one of the first to acquire a thorough knowledge of the AIC in South Africa, divided the second category mentioned above, the AIC, into two further sub-categories: the Ethiopian group and the Zionist group.

The Ethiopian group, according to Sundkler's view, are those independent indigenous churches which can be compared with the churches with Western connections, but which do not have any ties or connections with the West. The Zionist group, in Sundkler's opinion, consists of churches with a pentecostalistic approach and character. These churches are also sometimes styled Spirit churches or Prophet churches. They not only differ markedly from the African churches with Western connections, but also from the Ethiopian churches.

While research fieldwork was being done in Senekal in the Orange Free State, it transpired that members of the AIC themselves consider it important to distinguish between Ethiopian churches and Zionist churches. In Sotho they call the CWC Dikereke tsa molao, the churches of the law. By this they mean that these churches do not give much room for the movement of the Spirit. Not only their worship services, but also the ways the church is run, are governed by the church polity and regulations.

Against the churches of the law, the Dikereke tsa Moya, the churches of the Spirit, position themselves, These are the churches which make room for the Spirit, and allow the Spirit to guide, to take over and to reveal as the Spirit wills. Rules, regulations and liturgy, all these are subordinate to the Spirit. In fact, the liturgy is arranged in such a manner that scope is given for the Spirit to take over. That is why in the worship services of these churches the people dance and run in circles while accompanied by a musical instrument which could be anything from a drum to a dinner bell. This they do until they drop down in ecstasy. For them these are signs that the Spirit took complete possession.

Both interesting and significant is the fact that very often during the research fieldwork the Ethiopian churches were seen by the respondents as Dikereke tsa molao. This supports Sundkler's view that there is an important difference between the Zionist churches and the Ethiopian churches. Because the Ethiopian churches often follow the pattern of the CWC they were like the CWC classified by the respondents as churches of the law. The reason is clear. Both these groups are seen to be less open to the operations of the Holy Spirit, but are being governed by fixed rules and regulations in the church and by a prescribed liturgy in the worship service.

Turner classifies the AIC under two broad headings: Prophet-healing churches and Ethiopian churches. The former he sub-divides into Messianic, Revelation and Therapeutic churches, and the latter into Ecclesiastical and Evangelical churches. Turner does, however, admit that in South Africa there are many examples of churches which do not fit a specific category, but which are really a mixture of several different kinds (Turner 1968:32-33).

The ZCC is an example of a large church which does not completely fit the classification of either Sundkler or Turner. Its traditional customs of ancestral veneration and sacrifices and its somewhat magical view of the Bible and the Lord's Supper do give reason to place this church near the categories listed as Neo-paganistic and Hebraistic. Following Turner's classification, the ZCC fall partially under the Prophet-healing churches, but it rather belongs to the Messianic and Revelation church categories. (Hanekom 1975:97, Kruger 1983:7). The ZCC sees an intimate connection between the Lord's Supper and the Old Testament passover. In Ex 23:14-15 we read how God told the people through Moses to make a journey thrice a year to celebrate the feast of the unleavened bread. That is why the ZCC celebrates the Lord's Supper three times a year and only at Moria (Kruger 1983:7).

Furthermore, while the ZCC venerates its leader almost to the point of deification, this church also strongly reveals characteristics of a Messianic church. The leader occupies such an important role that he often assumes the place of Jesus Christ. Although assuming this role is not according to the teaching of the church, very many church members have such a perception of their leader (cf. Hanekom 1975:62-75). Martin (1964:122ff) supports this view. In her discussion of the Messianic churches in South Africa, she explicitly pictures Lekganyane, leader of the ZCC, as an African messiah.

Kruger (1983:11-12), in an article in which he evaluates the ZCC according to standards found in the Bible and pronounced by the Reformation, he concludes that "the ZCC must be regarded as an organization which misses the marks of the true church". Yet, he sees as positive the fact that the Word of

God is proclaimed and heard in the ZCC. This is also how Calvin felt about the Roman Catholic Church of his times, that even there God was still at work.

The fact that one of the biggest and best-known indigenous churches in Africa is characterized in such terms, again raises the question: Is a Reformed indigenous church in Africa a possibility?

This bring us back to the matter of how the indigenous religious groups see themselves. Fieldwork research among the Southern Sotho yielded noteworthy results. Hundred per cent of the respondents of thirty-three indigenous religious groups view themselves as churches and call themselves "church". Some of them were unhappy that they were often classified as African churches and movements. The Rev S A Mohono, the present chief leader of the Moshoeshoe Berean Bible Readers Church (MBBRC) and son of one of the foundation members of this church was particularly annoyed. He felt insulted because his church and others, as objects of study, were referred to as "church movements" or even "syncretistic movements". This would imply that these churches were seen as movements having a veneer of Christianity while adhering to African traditional customs and practices. Such a view virtually excludes the possibility that an African Independent Church of Reformed conviction could be found. However, research on the Moshoeshoe Berean Bible Readers Church opened interesting perspectives.

The Moshoeshoe Berean Bible Readers Church (MBBRC)

The MBBRC was established by Walter Matitta in Lesotho in 1922 after he had severed his ties with the Paris Evangelical Mission Society. Matitta was a Spirit-filled evangelist whose only desire was to preach Jesus Christ. Hundreds of people became converted during his ministry. Many people were also healed through the powerful prayers of Matitta. His purpose was not to establish a church, but when conflict arose involving the Mission Society, he and hundred of his followers left the Mission Society to form the MBBRC, an independent indigenous African church.

Subsequently it was stated that this church had dissociated itself from the churches with Western ties so as to enable the Christians to find their African identity. The purpose was, as stated in the constitution, to promote the faith and religion of the Basotho, and to teach the African people to hold on to the expectation that Jesus the King would return.

From these statements it is clear that the MBBRC desires to be an African Church for Africans.

Research revealed that the MBBRC is an Ethiopian type of church, but with a Reformed confession (Van der Spuy: 1990). But the church is not only based on a Reformed confession, its difference is further expressed in the daily living of its members. Through interviews it became manifest that the members of this church do not participate in the ancestral veneration, not in the worship services, neither privately at home. According to them the Bible teaches that the Holy Spirit is the only spirit through whom man's wants and woes can be made known to God the Father. They do not need the spirits of the deceased ancestors to convey the requests to God.

For any church a written confession is a basis document reflecting the standard of faith. To undergird the statement that the MBBRC is an African indigenous church standing on and operating from a Reformed basis, the confession of this church needs to be looked at in some detail.

The confession reveals similarities with the Belgic Confession, a document of Reformed origin. As a result of the process of indigenization, the Belgic Confession was not adapted in its totality, but underwent some editing in the form of abridgement or combination of articles.

The fact that the Belgic Confession is used as the basis of the confession of the MBBRC suggests that the leaders possessed some knowledge of the Reformed teachings. This can be explained by the fact that Walter Matitta, the charismatic leader, used to be an adherent of the Dutch Reformed Church in

Africa in the Orange Free State. After that he became a member of the Lesotho Evangelical Church at Peka in Lesotho. This church was established under the supervision of the Paris Evangelical Mission Society. In this church Matitta worked as evangelist and evangelical preacher.

One of the other foundation members of the MBBRC was Raymond Molupe Mohono of Berea, Lesotho. He must have possessed knowledge of the Reformed doctrine. In 1990 the writer of this article found on Mohono's premises, where his son and present leader of the church now stays, evidence of such knowledge. Among the discards he found a book on Reformed theology entitled: *In understanding be men - A handbook on Christian doctrine for non-theological students*. The author was the Rev T C Hammond of Trinity College, Dublin. This book must have been in the possession of Mohono during or shortly after the foundation of the new church.

These two people with their Reformed background were the leading figures in the compilation of the constitution and accompanying confession of the MBBRC.

In the introduction to the constitution it is stated that the intention of the church is to be a church of Africa which will lead people to perseverance in the true faith in Jesus Christ. In a following paragraph, under the heading, "The basis of the church", Zech 4:6 is quoted. This verse is then understood to assert that the church of God does not originate from the power of kings, but through the power of the Holy Spirit (MBBRC, Book 1. Basis of the church).

Today this church has no formal ties with Western churches, neither is there any Western leader in their midst. This church is proud to be a self-supporting and independent church of Africa.

Evidence from the confession of the MBBRC

The following free translation of the confession, which is only available in Sotho, can serve as evidence that the MBBRC as an African independent church is rooted in the Reformed faith:

1. We believe with the heart and confess with the mouth that there is one God in Spirit. He has no beginning and no end. He is immutable. He has no body. He is perfect wisdom, justice, truth and beauty. He is the Almighty God, the fountain of all visible and invisible.

2. We believe that the Spirit of the prophets is the Spirit of God, as the Bible itself witnesses. The same Spirit was also in the apostles. They described what the Spirit had revealed to them in order to preserve that for the generations to come, and to teach them.

3. We believe in the holy scriptures of the Old and New Testament as the Word of God.

(i) Five books of Moses: Genesis, Exodus, Leviticus, Numbers, Deuteronomy, Josua, Judges, Ruth, 1 and 2 Sam, 1 and 2 Kings, 1 and 2 Chronicles, Esra, Nehemiah, Esther, Job and Psalms.
Three books of Solomon.
Four books of the major prophets: Isaiah, Jeremiah, Ezech, Daniel.
Twelve minor prophets: Hosea, Joel, Amos, Obadiah, Jonah, Micha, Nahum, Habbakkuk, Zephania, Haggai, Zechariah and Malachi.

(ii) Four gospels and the book of Acts of the apostles.
Fourteen letters of the apostle Paul.
The seven epistles: James, 1 and 2 Peter, 1,2 and 3 John, Judas and Revelation.

4. We believe all the laws and decrees of God of heaven are written in these books.

5. We believe in the triune God on the throne, the Father, the Son and the Holy Spirit. These three are one God. Gen 1:26.27: Then God said, "Let us make man in our image". Gen 3:22: Then the Lord said, "Behold, the man has become one of us, knowing good and evil". These verses give us to understand that there is not only one person sitting on the throne of God, but it is one person speaking to more than one.

The New Testament teaches us about the Son and the Holy Spirit. Mark 1:11, Matth 28:19, Luke 1:35.

6. We believe that Jesus is the only Son of God; He is the Word who became flesh. All things came into being through him. John 1:1,3.

7. We believe that the Holy Spirit together with the Father and the Son created everything. In the beginning the Spirit of God was moving over the waters.

8. We believe in God who created the angels together with the things of heaven. There are angels who rebelled against God and who became the enemies of God and the Holy Spirit. They are the angels who even today are harming and deceiving the church with a lying spirit.

9. We believe in the discipline emanating from God and directed at the angels of heaven and the creatures of heaven. There are also archangels, in other words, chief angels. The same oversight is given to people on earth. Thy are the creatures of God on earth.

10. We believe that the holy God created people without sin, but that man was deceived by the snake. In that way the true light of man was darkened. By the death of Christ man was brought back to his place by God.

11. We believe that the sin of Adam is being transmitted through man's nature to all people. The children drink this sin with the mother's milk. Sin changed the nature of man so that man has, no will to make contact with God. In the same way there are wicked angels who are called the angels of darkness.

12. We believe that God himself in Jesus Christ, his only Son, elected generations of people he loved in order to show his mercy in the flesh. Luke 12:32, Rom 8:33,35.

13. We believe that Jesus assumed human nature, but that this human nature did not destroy his divine nature. He is the Son of God. He assumed these two natures, nevertheless he was the one and only Son of God. He assumed these two natures in order to manifest God's love. As sinful man he was born from a women. He took the punishment of the sin of mankind upon him. In his divinity he laid down his life for us. He had no relatives among men, Is. 53.

14. We believe that man is renewed. This new life is obtained through faith in Jesus Christ, in the name of God the Father through the power of the Holy Spirit.

Through faith

(i) man receives justification

(ii) man's relationship with God is restored

(iii) man becomes an heir of eternal life.

15. We believe that all the sacrifices and services offered by Irael were fulfilled in Jesus Christ. The laws and commandments were handed down by prophets and apostles so that we can do the will of God.

16. We believe there is only one church of God. There are, however, various ways of serving God, just as there are various organs in the body.

17. We believe the church has a sacred government, that is the guidance of the Holy Spirit. Elders and deacons form the church council who must lead the church through the Word of God. They have the responsibility to shepherd, to teach and to discipline so that the people of the church be guided correctly in the paths of righteousness.

From this confession it becomes clear that some of the main tenets of the Reformed faith are expressed here. They are the total corruption of man and his complete inability to restore his broken relationship with God; inherited sin, election in Jesus Christ, justification by faith, regeneration by God, participation in the new life through faith in Jesus Christ and through the power of the Holy Spirit. Other basic teachings held by the Reformed churches, and also found in this constitution, are the following: the Trinity, and the Trinity's role in creation and recreation, the two natures of Christ, the vicarious death of Christ, the inspiration of Scripture and salvation only through faith in Jesus Christ. It has to be noted, however, that Reformed dogmatics does not assert that Jesus was born as "sinful man" or that he "in his divinity" laid down his life as article 13 of the confession of the MBBRC states. Apart from this, the confession does produce ample evidence that the confession of the MBBRC reflects a Reformed character.

Evidence from the constitution

Further evidence as to the Reformed character of the MBBRC can be gleaned from its constitution. Under the title Sacraments, the two sacraments recognized by the Reformed churches, namely the Lord's Supper and baptism, are mentioned. But then a deviation follows. Marriage is seen as a third sacrament. Despite this addition, the chief leader, the Rev S A Mahono, is strongly convinced of the Reformed disposition of their church, but he sees his church as representing a second line.

Article one under the heading Sacraments states that the sacraments were instituted by the Lord to enable us to get a better understanding of faith, to display Christ's mercy and to strengthen our faith. Nowhere is the Roman Catholic idea to be found that Christ is sacrificed again. Also for this reason it can be argued that the recognition of marriage as a third sacrament does not signal a leaning to Roman Catholic doctrine, but should rather be seen as part of an independent and indigenous African theology.

Article two asserts that baptism is the sign that the sins have been washed by the blood of Christ.

A Bible study document together with reports regarding a conference held in 1987 came into possession of the writer. In this document where the Lord's Supper is discussed at length, it is taught that the blessing of this sacrament and of baptism is meant "for us and our children".

Concerning the liturgy and the administering of the Lord's Supper the document advises that the Lord's Supper be celebrated in the same manner as is customary in the Lesotho Evangelical Church or the Dutch Reformed Church of South Africa.

Conclusion

Reflecting on the confession of the MBBRC, on the excerpts from the constitution and on the views expressed in the conference documents regarding liturgy and the sacraments, one cannot but admit that the existence of an independent indigenous African church with a Reformed character is indeed a possibility. One can even go so far as to postulate that such a possibility became a reality in the MBBRC. The evidence adduced above together with the Rev S A Mohono's assertion that the church is essentially reformed, renders support for this postulation.

The evidence presented also urges one to be careful in making pronouncements regarding the AIC. As wide as the spectrum of belief of the mainline church is, so wide also is the spectrum of the AIC. The classification system outlined above shows that generalizations regarding the AIC are just as unacceptable as in the case of mainline churches which have fallen victim to unsubstantiated pronouncements.

The hope is being expressed that this essay would assist in resisting the tendency to summarily discard or denigrate the AIC. More positively, it is hoped that the research results would open people's eyes to look in a well-balanced manner to other forms in which the gospel can effectively be presented in Africa.

BIBLIOGRAPHY

Cronje, J 1990. Die Sioniste het ywer sonder kennis. *Die Voorligter* 53, 5 20-23.

Damane, M 1985. Walter Matitta (1885 - 1985). Paper presented at the National University of Lesotho.

Gilliland, D S 1986. How christian are African independent churches? *Missiology* 14, 259-272.

Hanekom, C 1975, *Krisis en Kultus. Geloofsopvattinge en seremonies binne 'n swart kerk.* Kaapstad: Academia.

Kruger, M A 1983. The Zion Christian Church in the light of the understanding of Scripture by the Reformation. *In die Skriflig*, 66, 2-12.

Martin, M L 1964. *The Biblical concept of messianism and messianism in Southern Africa*. Morija: Morija Book Depot.

Mohono, R M 1988. *Lengosa la Molimo Walter Matitta*. Morija: Morija Sesuto Book Depot.

Mohono, S A 1987. Spiritual maturity of Matitta and schisms. Paper presented at NERMIC Conference, University of Witwatersrand.

Moshoeshoe Berea Bible Readers Church, 1922. *The constitution of the Moshoeshoe Berean Bible Readers Church*. Morija: Morija Book Depot.

Sundkler, B G 1948. *Bantu prophets of South Africa*. London: Lutterworth Press.

Turner, H W 1968. A typology for African religious movements. *Journal of religion in Africa* 1, 1-34.

Van der Spuy, R J R 1990. Preliminary thoughts on the MBBRC. Paper presented at NERMIC Conference, University of Witwatersrand.

Van Wyk, J H 1973. *Die separatisme en inheemse kerklike bewegings onder die Bantoe van die Sothogroep*. Vol 1 en 2. Pretoria: RGN verslag.

PART TWO

ON HEALING IN AFRO-CHRISTIAN CONTEXT

CHAPTER 7
LEAVES OF HEALING
CHRISTIAN MISSION IN THE CONTEXT OF AFRICAN WITCHCRAFT

by

H-J Becken

"Leaves of Healing" was the famous monthly mission magazine edited since 1896 by John Alexander Dowie (1847-1907) of the Christian Catholic Apostolic Church in Zion, Illinois, U.S.A. This magazine was widely spread and read in African Independent Churches of the Zionist type and had a great influence on the development of their theology of healing in Southern Africa. Its title referred to the passage in Revelation 22:2, where John sees next to the river of the water of life also

> "the tree of life with its twelve kinds of fruit,
> yielding its fruit each month, and the leaves of
> the tree were for the healing of the nations".

When visiting Zionist Churches in Zululand and Natal, I sometimes found copies of this paper in the homes of bishops and prophets who cherished them for their holistic interpretation of the Biblical message so congenial with their own reception of the Word of God. Here they found a confirmation of their African missionary approach after having been disappointed by the established churches with their Western orientation of theology. They had left these churches. Nevertheless, they were thankful to them for bringing and translating the Bible into their mother tongue, thereby enabling them to find the Lord Jesus

Christ, the Saviour from sin and diseases, who was sent by God to free them from the powers of darkness surrounding them in their daily lives.

Strengthened by this ecumenical support, the Zionist churches went on baptizing the converts in the living waters of Zululand, casting out demons, speaking in new tongues, picking up snakes, and praying for the sick by laying their hands on them, thereby communicating faith in Jesus Christ to ever growing numbers of people (Mark 16,14-18). The African Independent Churches even today spread like a bush fire in this country and have proved to be a reformation movement in the context of Africa.

Professor G C Oosthuizen recently covered this subject in a scholarly way in his publication, *The Healer-prophet in Afro-Christian Churches*, Leiden: Brill, 1992. It would, therefore, be bringing owls to Athens should I attempt to write more on this subject. Rather shall I draw attention to the fact that this concern for a contextual African theology is also felt within the established churches. I shall dwell on two outstanding African church leaders who, in spite of being criticized and harassed by their church authorities, were led to initiate prayer healing movements in which the (not only) African perception of demons and of people using evil spirits to harm other people (let me call them "witches") are taken seriously in communicating the message of salvation in Jesus Christ.

One of them is the former Roman Catholic archbishop of Lusaka (Zambia), Emmanuel Milingo, who established a charismatic group named "Divine Providence Community" in Zambia. The other one is the former Lutheran bishop of the southern diocese of the Evangelical Lutheran Church in Tanzania. His movement of prayer healing groups is the "World Community Healing Revelations" with the centre at Njombe (Tanzania). The name of this man is Dr Judah B M Kiwovele. Both of them were ordained around the same time: Kiwovele in 1957 as pastor, Milingo in 1958 as priest. Both of them were guided by revelations of divine spirits to take up the struggle against the powers of darkness and to resort to healing prayers: Milingo in 1973 in his urban environment at Lusaka; Kiwovele only in 1985 in his remote rural setting in the South-West Highlands of Njombe, on the mainland of Tanzania. Milingo was able to follow his healing ministry for nine years, until he was summoned to Rome where his case was investigated. Later he was given an appointment as Special Delegate to the Pontifical Commission for Migration and Tourism in 1982. Kiwovele was only tolerated by his diocese for two years, until he was sent to retirement as a bishop in 1987, and the members of the diocese were forbidden by pain of excommunication to seek services by the Healing Revelations Community. However, the Lutheran Church on a national level did not yet find a way how to deal with this movement, and Kiwovele since then conducts his healing ministry outside the church.

The phenomenon to which these Christian healers responded, is the same experience often encountered in the healing services of Zionist and Prophetic Independent Churches, namely that strange voices speak out of the patient. The indwelling "demons" frequently confess to have been sent by an identified person to harm their victim by a disease in order to destroy him/her. After this confession, the evil spirit (demon) will be driven out of the sick person in the name of Jesus Christ. The demon will either return to the witch who sent him or he will be annihilated, while the patient is healed. This exorcism, Milingo calls it the "service of delivery", has a long tradition in the Christian churches, not as a magical act, but as an effective prayer for delivery from evil. This practice in the churches of the West faded away in the course of the last centuries.

It is a useless question to ask whether demons exist and evil people are able to make use of them to harm other people by sending them to inflict diseases on them. For people who experience this phenomenon, this is a reality. Missionaries are blamed to have kept a blind eye to these very existential sufferings of their African converts. Milingo says that missionaries introduced every kind of indigenization to the church in Africa, using drums and African music in worship services, but described, on the other hand, everything African as superstition or psychic deformation. They never accepted in their healing ministry the possession by evil spirits as a disease worth healing, since they called it, by their own name, a hysterical or psychosomatic disease. They, therefore, considered anyone engaged in fighting it as an imbecile who chased the wind. Kiwovele points out that this superficial approach resulted in secret survival of witchcraft within the churches, even in their leadership. He refers to well-known events in the pioneer missionary period, when ignorant missionaries protected the lives of confessed witches from the heathen chiefs by granting them shelter in the mission stations, thereby allowing them to continue their practices under the cloak of being members of their churches. On the other hand, these Christian healers also experience the guidance and support of corresponding good spirits of God. Milingo terms them "the world in between". This is also the title of his book.

The two gifts - prophecy and casting out of devils - are gifts of God to the community, communicated through ordinary human beings. Therefore, besides individual counselling, the praying team is of the greatest importance, and in this team "the world in between" joins forces with its earthly members, who ought to be true followers of Christ and who are prepared to accept sufferings in the imitation of the Lord. Besides his awareness of the presence and interceding activities of angels, Milingo - from his Roman Catholic background - also relied on the assistance of the saints. Many of them had experienced diseases, and there was very little need to explain to them the problems of the sick, e.g. epilepsy, leukaemia, asthma and continuous headache. Their favourite saints were Mother Mary, St Joseph, St Andrew Bobola, St Peter and St Paul, St John Vianney (the Curé of Ars) and, of course, the African St Patricia (from Kenya).

Kiwovele received a revelation in 1985, when God called him as a prophet, and told him that his only purpose was prayer and intercession for the sick, and that healing revelations would spread all over the earth bringing love, responsibility, peace, happiness and harmony in the creation. People would renounce sin and the devil and his angels who used to control their lives. Kiwovele feels the presence of and listens to the advice of intermediaries to God. He terms them angels, messengers and holy spirits. Through revelation, he knows a number of angels by name, e.g. Mungu Mfadlili (Graceful God), Mungu Nguvu (Powerful God), Faraja (Comfort), Furaha (Happiness), Upendo (Love), Huruma (Mercy), Asifiwe (Praise the Lord).

This notion of angels and saints has been alive throughout the centuries of Christian church history. Even if theologians may be hesitant today to speak about them, popular Christianity sticks to them. However, it is the merit of African theology to interpret them not only as kind guardian spirits, but also as warriors fighting, together with Christ, the powers of evil as depicted in the book of Revelation in the New Testament. The healing teams hear these angels speaking and praying, and they experience that they quickly disturb the presence of the evil spirits among the patients. In the Healing Revelations Community they also give messages which are recorded literally together with the dates on which they were given.

The discovery of the gift of healing came to both African church leaders not by theological studies but in the course of their normal counselling activities as pastors. On 2 April 1973 a woman who had been treated without results in a mental hospital, came to Milingo's office and explained that she had suffered for five months without eating; she only drank water or soft drinks. They prayed together. A few days later she came back and told him her whole story: She feared her child because she did not think he was a human being, and she constantly heard voices speaking to her. Even after hearing confession and celebrating mass, these symptoms continued to trouble her. Then suddenly the idea glowed in his mind to look into her eyes, to treat her with the sign of the cross, and to speak to her soul. He followed this instruction. The Lord delivered her, after the offering of prayer, both in body and soul. She recovered from her mystical disease and could eat again, she no longer heard voices and no longer feared her child. Thereafter, more demon-possessed people approached him for help. In the course of time, Milingo understood his mission as one of delivering from diseases caused by demons and witches.

Kiwovele had an experience of this kind in his own family when he was informed during the session of his diocesan assembly that his eldest daughter Rehema, who lived with her family at Dar-es-Salaam, had a difficult delivery which almost claimed her life and that of her baby. When her brother Justin and a few friends prayed for her, demons responded out of her mouth. She returned home to Njombe. A few days later a young lady by the name of Florensina, who lived in the bishop's family, fell sick and also showed demonic symptoms. Rehema was healed on 26 April 1986 through prayers in the name of Jesus Christ. God's messenger confirmed her healing in the evening of the following day. Florensina was also healed a few days later. Both ladies received spiritual gifts of visions and of healing by prayer. Kiwovele regards this experience as the beginning of Healing Revelations. Healing teams then began to visit sick people, not only in the congregation, but also in the broader community. The demons in sick people were no longer able to hide before the angelic spirits. By that time this ministry was performed with the consent of the diocesan assembly.

Both healers emphasize that all healing is an act of God. Heavenly spirits and human beings can only pray for God's healing, who will then respond at His will and at His time. And it is God alone who can remove visible disabilities and cast out or destroy evil spirits and demons, thereby granting holistic healing. Prayer is not a magical rite, but rather an intercession by faith in the name of Jesus Christ. Some people had already been healed during the hymn before actual prayers were said. Sometimes the mention of the name of Jesus Christ resulted in casting out of demons. Other people took months or years before they were healed. Some sick people gave up hope and never regained their health. Commonly, the leader of the team lays hands on the sick person, while Christian and heavenly spirits join in free prayers. Then frequently the Holy Spirit or an angel will diagnose the disease of the patient. Milingo reports that this is the most sorrowful part of the healing session, because the sick are made aware of their suffering. Epileptics then fall into convulsions, tumors become more painful, wounds are burning like fire, possessed people cry and shake, twist themselves and roll on the ground, and those troubled by anxieties and worries sometimes just weep. Kiwovele points out that this diagnosis by the spirits reveals the demonic power behind the disease. It also happens that no demon is detected; then it is a disease of purely natural causes, a normal medical case. Witchcraft can also be detected by this diagnosis; then the demons confess who sent them and how it was done. However, Kiwovele stopped this public diagnosis in 1988 when he observed that some people tried to prepare and to execute revenge, which would again destroy their holistic health. In the prayer services of the Healing Revelations, pastors and evangelists are also preaching to communicate faith in God through Jesus Christ. In this process the Old Testament is used as a stepping stone to the understanding of the New Testament. Sometimes angelic spirits preach in a very contextual way, because they also see the secrets of families, especially concerning human-controlled evil mystical powers. Milingo also sprinkles holy water and raises the crucifix "to seal the work of Christ by His precious blood." A prayer of thanksgiving and a special blessing to the healed person ends the service.

As we have observed, the two personalities introduced here came from different church denominations, worked in different social settings and among different nations, and at different times. While their basic experiences with evil and heavenly powers in the spirit world had been similar, these contextual factors have shaped their ministry along different lines. There is also no indication that they knew each other. However, one observation is striking, namely that in both of their geographical regions, there was a blank spot on the map of African Independent Churches. To quote David B Barrett, "In contrast to its neighbours Kenya and Uganda, Tanzania has never had large-scale independent church movements". (*Schism and Renewal in Africa*, O.U.P. 1968). Missologists assume that this was due to the influence of Bruno Gutmann (1876-1966) who developed a profound understanding of the meeting of gospel and African societies and of the real scheme of relationships by which the existence of people is determined. Zambia had its great time of African Independent Churches, when Regina Alice Lenshina gathered her Lumpa Church in 1954. However, ten years later, this movement was crushed by the new independent government, and since then a similar vacuum as in Tanzania existed when Milingo began his ministry of delivery in 1973.

This observation leads to the conclusion that the non-existence of African Independent Churches in these countries built up a pressure in African spirituality which found an outlet through the valve of the above-mentioned church leaders and healing prayer movements.

It is deplorable that the two established churches concerned were unable, for whatever reason, to integrate these healing movements and to give way to the African Reformation. However, the lives of these men are "leaves of healing", leaves of the tree of life for the healing of the nations in Africa.

Besides personal interviews, I acknowledge the use of two sources in compiling this article, viz.

Milingo, E 1984. *The World in Between. Christian Healing and the Struggle for Spiritual Survival.* London: Hurst.

Kiwovele, J J. The Spiritual Warfare: Healing Revelations in Njombe, Tanzania. (Not yet published).

CHAPTER 8
TRADITIONAL HEALING IN SOUTH AFRICA

by

S.D. Edwards

Introduction

Spiritual and community healing dances can be observed throughout South Africa as members of African Indigenous Church (AIC) groups participate in healing that is both perennial and renewing, traditional and transformational.

It is appropriate to start with an example of a traditional healing dance by the "first" or "real" people, as the Kalahari !Kung refer to themselves, as described below by Katz and Wexler (1989:23).

"!Kung healing involves health and growth on physical, psychological, social and spiritual levels; it affects the individual, the group, the surrounding environment and the cosmos. Healing is an integrating and enhancing force, far more fundamental than simple curing or the application of medicine.

Sometimes, as often as four times a month, the women sit around the fire, singing and rhythmically clapping as night falls, signalling the start of a healing dance. The entire camp participates. The men, sometimes joined by the women, dance around the singers. As the dance intensifies, n/um ("energy") is activated in those who are healers, most of whom are among the dancing men. As n/um intensifies in the healers, they experience !kia ("a form of enhanced consciousness") during which they heal everyone at the dance. The dance usually extends far into the night, often ending as the sun rises the next morning. Those at the dance confront the uncertainties of their existence and reaffirm the spiritual dimension of their daily lives".

The Universality of Healing

As we are instructed in the example of !Kung healing, it is vital to remember, in the first instance, that all healing is universal, transcultural and perennial. We are concerned with universal human healing of man and for man. As a specific form of helping, healing can be traced to the beginnings of recorded human history.

To heal means to make whole. Healing implies a transformation from illness towards health. When we speak of traditional healing, we are concerned with perennial healing approaches that have been passed on over generations, often in the form of oral transmission of knowledge, beliefs and practices that serve to both preserve individual, family and community homoeostasis and to transform individuals and societies towards health.

The psychology of traditional healing is of an original and universal form, as conveyed by the Greek concept of psyche or the African concept of umoya (Nguni) or moya (Sotho), i.e. breath, soul or spirit as revealed in various forms, natural, human and supernatural. The essential theme is one of psychology as revealed in helping, healing human relations within the spiritual and community context, a pure psychology relatively undiluted by artificial academic distinctions, schools of thought or narrowed professional interests. "I am because we are" (Mbiti 1970) is the existential and phenomenological theme.

Healing is a special way of helping. It has ancient and proud traditions. More specific terms such as medicine, counselling and psychotherapy are all synonyms for the broader and more essential term healing. The term traditional healing is often used synonymously with more specific terms such as traditional medicine, folk healing, indigenous healing, allopathy, prescientific medicine, alternative medicine, etc. We should remember the common denominator of essential and universal healing when such terms are used.

One reason for the use of various terms is that illness and healing are vital matters of interest to people from various walks of life, persuasions and professions. The study of forms of traditional healing is an academic and professional multidisciplinary meeting ground for traditional healers, medical doctors and specialists: especially psychiatrists, psychologists, social workers, nurses, medical anthropologists, medical sociologists, primary health care workers, etc., who also attribute different meanings and labels to the phenomena under investigation. Even the same individual can perceive the same phenomenon in changing and different ways, for example, a traditional healer who is also a marine biologist may view water as life sustaining (if extremely thirsty), as healing or holy water (e.g. isiwasho) or H^20 in different instances.

There are many universal characteristics of healing. Some more original and essential characteristics based on the work of various authors are outlined below (Torrey 1972; Wolberg 1982; Rogers 1980; Cheetham and Griffiths 1978; Brammer and Shostrum 1982; Corsini 1984; Edwards 1988; Katz and Wexler 1989) are outlined below.

* The help of Mr. B.P. Mosisidi, in compiling Sotho equivalent translations of various terms and of Mr. M.M. Hlongwane for his fruitful discussions, is gratefully acknowledged.

(a) Healing dialogue. The essence of the healing relationship lies in the interview with its characteristically human helping dialogue.

(b) Spirituality. Healing involves a beyond-the-self dimension, for example God and revered ancestors are experienced as present in healing.

(c) Healing power. Throughout the world there is recognition of such a traditional power, called n/um by the !Kung, mana by the Figians, Wakan by the Nkota and Chi by Chinese Taoists.

(d) The healing exchange. There is always a give and take to and from the healer(s) and person(s) being healed. In taking on the illness, the healer is giving, in giving up the illness, the ill persons are taking.

(e) Community. Healing is a resource to be shared, which expands and is accessible throughout the community as in the ritual dance of the !Kung or Zionist prayer meetings.

(f) Shared world view. This includes the aura and atmosphere of the healing situation made closer through a shared language and culture.

(g) Healer qualities/characteristics. Authentic, genuine care and empathy/understanding make the relationship acceptable to the ill and therapeutic.

(h) Expectations of the ill person(s). Trust, hope, respect and faith in the healer help to heal the illness particularly if accompanied by a healer who listens and accurately divines/defines the problem, strengthens expectations and allows healing to take its course.

(i) Emotional arousal. The facilitation of emotional arousal in an intense relationship brings about attitudinal, behaviourial and cognitive change.

(j) Information. The offering of new information concerning the illness, typically leads to a consideration of alternatives, choices, decisions and problem solving.

(k) Learning and change in cognition, behaviour and affect all help in healing.

(l) Approach. The particular approach to healing or mythical rationale within which the healing takes place typically includes a reason for the illness, compatible with the shared world view and provides reassurance and healing in itself.

(m) Methods and techniques of healing. There are a great variety of healing methods and techniques, from ritual dance to pharmaceutical preparations.

(n) Holistic healing. We need to emphasize the essentially holistic nature of healing which includes all the above characteristics and more.

While on the surface there may seem great variation in people, views and healing in South Africa, core universal and essential components exist which are shared by many people across the globe. What follows is a synopsis of some traditional views on illness and healing in South Africa as particularly relevant to indigenous people of Africa. Similar views are held by people and traditional healers of European, Asiatic, Chinese and other ancestry and descent. The decision to focus on the views of indigenous African people is because these views form the majority of views of people in South Africa. Views are described in terms of the two main language groupings: Nguni (Zulu, Xhosa, Ndebele and Siswati) and Sotho (Tswana, Sesotho and Sepedi), hyphenated in that order.

Some traditional views on illness and healing in South Africa

Traditional African views on illness and health in general, and mental illness and health in particular are holistic and cosmological in emphasis. Traditional beliefs and practices concerning illness and health are still widely followed, especially in rural areas of South Africa. This is understandable in view of the fact that these beliefs and practices form a coherent system that has maintained individual and social equilibrium for generations. Secondly, more expensive modern medicine is not as readily available, particularly in rural areas.

Central to an understanding of these beliefs and practices is an understanding of traditional religion, which embodies the essence of all religions in its reference for the First Creator God, uMvelinqangi/Mmopi through elder kinsmen, both living and dead. This African spiritual component of being has been eloquently described by Holdstock (1981:128):

> "The relationship with the ancestors and through the ancestors with God permeates all being".

> Ancestral reverence is a primary factor associated with continued good health.

Sorcery or ubuthakati/boloi, on the other hand refers ultimately to "the manipulation and expression of anger and the desire to destroy" (Berglund 1976:295). It represents all the forces of evil and illness. With the concept of thanatos, Freud has similarly given recognition to human destructiveness and violence as a most basic cause of all illness.

Traditional African religious and magical theories of illness and health are inextricably interlinked as evident in the popular beliefs that man is most vulnerable to sorcery once for some reason the ancestors are "facing away" (Berglund 1976), and should a man then strengthen himself and his family by performing appropriate rituals to the ancestors, this will ensure continued good health.

Murdock et al. (1980) have developed a comprehensive classification of theories of illness in underdeveloped societies throughout the world. They make a basic distinction between theories of natural and supernatural causation which is similar to the umkhuhlane: ukufa kwabantu distinction made by Ngubane

(1977), working among the rural Nyuswa valley Zulu people near Durban. The Sotho equivalent is mokgohlane: lefu la batho.

Umkhuhlane/mokgohlane refers to the explanation of illness by natural causation. This category is recognized by modern medical science with its empirical traditions, e.g. as in the case of infection, stress, organic deterioration and accident (Murdock et al. 1980). Isithuthwane/sethwathwa or epilepsy and isifuba somoya/sefuba sa moya or asthma are relevantly classified in this category. Ukufa kwabantu/lefu la batho on the other hand, attributes illnesses to supernatural causation. The term literally refers to "disorders of the African people". To quote Ngubane (1977:24)

> "The name is used mainly because the philosophy of causality is based on African culture; this means not that the diseases or rather their symptoms, are seen as associated with African peoples only but that their interpretation is bound up with African ways of viewing health and disease."
>
> "Murdock et al. (1980) have divided their supernatural category into three basic divisions. These three divisions can best be conceptualised as three different traditionally acceptable attributions made by the afflicted to explain the affliction.

The traditional ukufa kwabantu/mokgohlane theories* can be subsumed within these divisions as follows:

a) Animistic theories ascribe the disorder to the behaviour of some personalized supernatural agent such as a spirit or God, for example:

   abaphansi basifulathele/ba fatshe ba re lahlile - withdrawal of protection of the ancestral shades, mostly caused by disharmony within the home.

ukulahla amasiko/ho lahla maseko - failure to perform necessary rituals such as sacrifices to the ancestral shades.

ukuthwasa/ho thwasa - a "creative illness" following the calling by the ancestral shades to become a diviner, a religious conversion experience.

ukudlula/ho tlola - failure to indulge in abstinence behaviour during a period of mourning believed to result in a form of compulsion neurosis.

izizwe/moya o ditshila - aggressive spirit possession occurring by chance.

indiki - another form of spirit possession from the north, especially Swaziland.

(b) Magical theories attribute the disorder to the covert action of a malicious human being who employs magical means to injure his victim, for example

ufufunyane/mafonfonyane - spirit possession attributed to sorcery.
idliso/sejeso - poisoning attributed to sorcery.

iqondo/leqotho - genito-urinary disorders attributed to sorcery.

umeqo/mohlala - disorder attributed to stepping over the harmful concoction of a sorcerer.

umhayizo/lehabeya - crying attacks attributed to sorcery by love potions.

* For a more detailed discussion of these theories the reader is referred to Conco (1972), Berglund (1976), Ngubane (1977), Edwards (1985) and Gumede (1990).

uvalo/letswalo - anxiety attributed to sorcery aimed at lowering the defences.

ibulawo/bolawo - bodily pains attributed to sorcery.

tokoloshe/thokolosi - witchcraft through a familiar, the supernatural agent of a witch.

c) Mystical theories - explain disorders in terms of an automatic consequence of some act or experience of the afflicted person, for example:

umnyama/bofifi - experiencing illness or adversity because of contact with places or people immediately associated with major life and death events, e.g. birth, death, sexual intercourse, menstruation.

umkhondo omubi/mohlala o mobe - a dangerous track or ecological health hazard such as lightning.

Such theories form a corner-stone of traditional African cosmological, religious, social and moral world views of good and evil, health and sickness.

Indigenous diagnosis and treatment of both umkuhlane/mokgohlane and ukufa kwabantu/lefu la batho categories of illness are typically and traditionally treated by traditional healers. The traditional healer has always been a person of great respect in the community, a medium with the ancestors (amadlozi/badimo) and the First Creator God (uMvelinqangi/Mmopi) (Ngubane 1977, West 1975, Cheetham and Griffiths 1982). Historically the traditional healer had a multifaceted role in the community including that of

- diviner/priest, accepted medium with the ancestral shades and first creator God, religious head of society, prominent at all major rituals, e.g. concerning healing, fertility, life and death.

- protector and provider of customs, sociocultural cohesion and transformation, legal arbiter at public divinations, ecologist, rainmaker, etc.

- specialist in preventive, promotive and therapeutic medicine including the use of traditional pharmacology (Edwards 1987).

A good case can be made for there being three broad overlapping categories of traditional healers in South Africa today, i.e. the traditional doctor (inyanga/ngaka), diviner (isangoma/leqekga) and faith healer (umthandazi/morapelli). The advent of the umthandazi/morapelli can be traced to the rise of the African Indigenous Church movement and it has been argued that many of the traditional roles of the isangoma/leqekga have been taken over by the umthandazi/morapelli (Lee 1969, West 1975). There are various specialists within these three broad categories of indigenous healers for example:

(i) inyanga yomhlabelo/ngaka e qobolang - the orthopaedic surgeon
(ii) inyanga yesisu/ngaka ya mala - specialist stomach doctor
(iii) isangoma sekhanda/senohe - the listening or head diviner
(iv) isangoma samathambo/selaodi - bone-throwing diviner

(v) isangoma sabalozi - whistling spirit diviner or ventriloquist

(vi) umthandazi wamaZiyoni/morapelli wa siyone - faith-healer of Zionist religion

(See Conco 1972, West 1975, Ngubane 1977, Edwards 1985, Oosthuizen 1986, Oosthuizen et al. 1989, and Gumede 1990 for other categories of healers).

These three broad categories of indigenous healers use a variety of healing methods, which will be discussed separately under the three categories.

(a) Inyanga/ngaka. These traditional doctors are usually male. To quote Ngubane (1977:102)

"A man who wants to be an inyanga (doctor") usually gets himself apprenticed to a practising inyanga for a period of not less than a year. At the end of his training he pays his master a cow or its equivalent in money usually not less than R20. Sometimes a doctor passes on his skill to his sons, who show an interest in medicine."

Izinyangas typically use herbal medicines (amakhambi/dipitsa) in the treatment of their patients. Refer to Ngubane (1977) and Gumede (1990) for detailed descriptions of these remedies. Some typical treatment methods are recorded below.

(i) isinyamazane/diphoofolo - the Western equivalent of an immunisation or preventative medicine - consists of fat, pieces of skin and feathers and is given to young children.

(ii) isihlambezo/sehlapetso - a liquid medicine taken by pregnant women over a period of time both as tonic and preventive medicine.

(iii) ukuphalaza/ho kapa - vomiting induced by an emetic to 'clear out' the system.

(iv) ukuchatha/ho tshela - enema or laxative.

(v) ukugcaba/ho phatsa - incisions made at the joints, believed to be the most vulnerable areas of the body.

(vi) ukugquma/ho arubela - steam bath to induce perspiration and reduce fever in adults.

(vii) ukuthoba/ho thoba - a fermenting treatment applied in such cases as rheumatic muscular pain or aching, swollen feet.

(viii) ukulumeka/ho lomeha - blood letting, e.g. for severe headaches.

Herbal medicines (amakhubalo/ditlhare) are also used in a ritual and symbolic context.

Such medicines are classified according to colour and administered in strict serial sequence and for example given at night, day-break, daylight.

(i) Imithi emnyama/meriana e metsho - black medicines

(ii) Imithi ebomvu/meriana e mefubedu - red medicines

(iii) Imithi emhlophe/meriana e mesweu - white medicines

Medicine typically used in symbolic ritual sequence according to colour are:

(i) Ubulawu/bolawo - a liquid medicine used as a black, red or white emetic.

(ii) Insizi/messesetso - powdered herbs or roots or animal medicine - which is always used as a black medicine - to withdraw illness.

(iii) Intelezi/lekgalana - liquid medicine used as a white medicine to render free from imperfections, e.g. after sickness is taken out by black and red medicine. (See Ngubane 1977:110).

(b) Isangoma/leqekga. The traditional diviner who is usually a woman, shares a comprehensive knowledge of medicine with the doctor (inyanga/ngaka). To quote (Ngubane 1977:102)

> "A person does not choose to become a diviner (isangoma), but is chosen by her ancestors, who bestow upon her clairvoyant powers. A neophyte learns about medicine from a qualified diviner to whom she is apprenticed for some time, but in addition some medicines are said to be revealed to her by her ancestors," (i.e. As she undergoes the ukuthwasa/ho thwasa process).

While there are different specialist diviners, the diviner typically diagnoses through her mediumship with the ancestral shades. In practice, the most common method is the vumisa technique, with the diviner telling the patient of his illness, and honing in on problem areas depending upon the degree of expressed agreement of the patient and/or his relatives. In cases of public disputes, an umhlahlo (public divination) with a higher grade diviner may be sought (e.g. isangoma sabalozi) in traditional rural settings.

Despite her comprehensive knowledge of medicine, the roles of the isangoma rates as diviner and as medical practitioner do not overlap. In the first place she is consulted as a diviner to ascertain the causes of problems. The specialist isangoma will often refer patients for medical treatment to another practitioner.

Umthandazi/morapelli literally means one who prays or a faith healer. West (1975) distinguishes between faith healers and prophets as quoted below:

"The faith-healer (probably the smallest group although no figures are available) is a Christian who may belong either to a mission or independent church. The power to heal is believed to come from God (although in some cases it may be thought to come from God indirectly through the shades) and a period of training as a healer may or may not have been necessary." (West 1975:96). "The prophet is a healer who is found mainly in the Zionist and Apostolic churches, who has the ability to predict, heal and divine, and who draws power to do this from God. Although in many cases the power may come from God through the more direct agency of certain guiding shades." (West 1975:98).

The strongest prophets usually heal through private consultations and most have a history of illness which has been cured by another prophet (West 1975).

West (1975) lists the main treatment methods in order of frequency of a typical prophet as follows:

(i)	Prayer	in 100% of cases
(ii)	Holy Water	in 79% of cases
(iii)	Baths	in 52% of cases
(iv)	Enemas	in 45% of cases
(v)	Steaming	in 42% of cases

The above is a very cursory outline of some indigenous views on illness and healing in South Africa. For more detail refer to Conco 1982, Ngubane 1977, Edwards 1985, Oosthuizen 1986, Oosthuizen et al 1989 and Gumede 1990.

It should also be emphasized that these views and patterns are continually changing and transforming society in a mutual reciprocal relationship (Edwards 1987). For example, the completed ukuthwasa/ho thwasa process may be described as a crisis in living (Kruger 1974) accompanied or followed by a creative illness (Buhrmann 1977) in which there is integration and resolution in the life world of the healer and the secular and religious community, who benefit from the healer's divinatory abilities as a result of the religious conversion experience. Such religious conversion experiences occur in various forms amongst all peoples constituting a marvellous way of humanity caring for its afflicted.

As far as the three broad categories of traditional healer are concerned the inyanga/ngaka has received the greatest recognition in terms of popularity, professionalization and recognition by modern medicine. Traditional herbal and other pharmaceutical preparations have received increasing scientific study leading to important medicinal discoveries. For example, reserpine from the rauwolfia root was introduced into modern Western psychiatry in the 1950s as a major tranquillizer. At the time it was recognized as having been in wide use in West Africa for centuries (Torrey 1972). Such very valuable medicinal discoveries are continuing at the moment (Hutchings 1992, Gumede 1990).

The changing roles of the isangoma/leqekga are apparent at various levels (Edwards 1987). Whereas this does not appear to have always been the case, sex roles have become more differentiated in that these practitioners are almost always female, reflecting, i.a., a rebellion against more repressed, traditional, female roles. There appears to be greater specialization and differentiation in the role of the diviner. Many diviners actively practise traditional medicine. Popular diviners are common, those of standing rare. The number of diviners is increasing for both health as well as social and economic needs, e.g. the lucrative demand for such services.

It has been observed that the faith healer (umthandazi/morapelli) has assumed many of the diviner's traditional roles. There has been a vast proliferation of African Indigenous Churches (AIC) of which the following of the Zionist religion form by far the largest religious denomination in Southern Africa. The AIC stem from a religious movement started in the last quarter of the nineteenth century in South Africa, eventually playing a role in the formation of the African National Congress (ANC) in 1912. As a whole the AIC are generally apolitical, with the prime focus being on healing and development (Oosthuizen 1986, Oosthuizen et al 1989). It is an interesting question whether any other organization is doing more for the health of the majority of South Africans than the African Indigenous Churches.

The theme of healing as transformation may be observed on various levels including the following:

(a) Holistic change of an essentially spiritual and community nature as in the ongoing healing event.

(b) Interpersonal change as in the changing healing dialogue of, for example, the vumisa/laodisa technique where the healer divines and the ill and/or relatives contractually agree on the illness and healing.

(c) Individual change where the healer is changed by the call to become a healer and other individuals later become healed by the healer.

(d) Healing method changes, e.g. through discovery of new medicinal herbs leading to biological and other changes.

(e) Social change, e.g. AIC caring for spiritual, social, political and economic needs, in what Katz and Wexler (1989) term a "renewed tradition".

Traditional and modern healing

Universal human healing is both a perennial and changing event, with modern healing having essentially grown out of traditional healing. Any distinction between traditional and modern healing is essentially an artificial and arbitrary one. Universal components are fundamentally far more important than relative superficial differences, whether the concern is with the healers, the ill, the healing event or health care systems.

Probably the most common synonym for traditional healing is traditional medicine which i.a. reflects the interests of established modern medicine in its own foundations, that is, traditional healing and medicine. The term medicine is used here in its broadest sense as a science and art of healing illness.

The distinction between traditional and modern medicine is internationally accepted as denoting more regional, indigenous, alternative, functionally strong approaches to medicine as distinguished from modern biomedically orientated structurally dominant systems (Edwards 1986). It is an arbitrary but useful distinction.

Problems in Southern African medicine are fundamentally the same as those in other countries where socio-economic, political, legal and other factors all form part of the total healing context. Problems in South African medicine have, however, been exacerbated by historical, racial, violent politico-legal and societal divisions, which have tended to maintain regional separation between communities and traditional and modern approaches to illness and healing (Edwards 1989). For this reason, in particular, the distinction between traditional and modern medicine remains very real.

A further, fundamental reason for the distinction between traditional and modern medicine is the relative natural and supernatural orientations of modern and traditional medicine respectively. This distinction itself reflects the situation of contemporary humanity, living in a technocratic and divided world which is characterized by ever accelerated change, violence and loss of spirituality.

Gumede (1990:153) has documented the following differences between traditional and modern healers:

(a) Modern healers are originally Western in origin. They are imported into Africa and are called medical practitioners or just doctors.

Traditional healers are indigenous and African in origin. They are called medicine men, herbalists, izinyanga, dingaka, igquira, etc. all of which mean traditional healer.

(b) Modern healing dates back to Hippocrates and the early Greek; traditional healers were in existence and practising as they do today in Kush/Ethiopia some 4 500 years ago.

(c) Modern healers are regarded as rational while traditional healers are described as irrational.

(d) Modern healing is described as scientific while traditional healers are regarded as empirical and unscientific.

(e) Surgicial procedures by modern healers are planned, scientific and based on the study of gross and morbid anatomy. Their operations are done under deep anaesthesia with high technology available to provide the back-up system. Surgical procedures by traditional healers are regarded as unscientific, crude and unplanned in the view of "modern" understanding of healing.

(f) The training of a modern healer takes some seven years after the final school year. The training is available to anyone who has the necessary means and entrance qualifications to enter a medical school. The training of a traditional healer is from father to son or from master inyanga to trainee, apprentice, journey man and full blown inyanga. It takes a lifetime. The constitution of the inyanga's national association prescribes a period of ten years (Article 7.1).

(g) In modern medicine, aetiology of disease is based on germ theory. Disease follows a breach in the laws of nature, e.g. a prick by a rusted dirty nail introduces germs like staphylococci which cause blood poisoning and abscess formation. While recognizing that imikhuhlane (colds) are natural, traditional healers, mainly regard disease as unnatural and typically man-made through the agency of a spirit. While the modern healer's problem is to find the offending organism and deal with it using pharmaceutical high technology now available to modern medicine, the traditional healer is not only expected to diagnose the malady from which the patient is suffering but should also be able to decide whether or not he or she has been bewitched - and by whom (without mentioning names).

(h) For modern healers diagnosis entails what (germ) caused the illness while for the traditional healer it is not only what illness ails the patient but also who caused the illness.

(i) The modern healer's treatment is specific, individualized and streamlined to meet the presenting problem. If the lungs are sick, go to the chest specialist, if a bone is broken, go to the orthopaedist who knows his carpentry, etc. The modern healer's approach is to heal the affected part. The traditional healer's approach is holistic. A person is a total being including body, mind and soul; therefore, healing is a total process involving the living and the dead, the "natural, preternatural and supernatural" in addition to the patient.

(j) The modern healer treats the disease; destroys the offending organism and the patient will be well. The traditional healer treats the patient within his environment - physical, spiritual, emotional, past and present.

(k) There are some 20,229 (1986) doctors registered with the South African Medical and Dental Council to minister to a guesstimate population of 35 million people in South Africa. It is estimated that there are about 120,000 traditional healers, that is izinyanga only, excluding izangoma, izanusi, abalozi. The correct figure is probably much higher.

(l) The code of conduct for modern healers is laid down by the South African Medical and Dental Council. Traditional healers belong to various associations.

(m) Stress is laid by modern healers on (a) the troubled, viz. the disease and (b) the seed, viz. the germ and the disease it causes, whereas traditional healers lay stress on the troubled, viz. the patient and the soil, i.e. the environment - physical and cosmic.

(n) The language of the modern healer is scientific in approach, aetiology, symptomatology, diagnosis, epidemiology, endemiology, curative, preventative, prognoses, rehabilitation, morbidity, mortality, etc. The idiom of approach of the traditional healers is social, political, economic, moral, religious and even recreational and change of environment, e.g. in hysteria.

Along with these differences, Gumede (1990:153. 231), who himself combines both approaches, observes that both modern and traditional healers have the same goal of helping the ill and healing illness and that what is little appreciated is that there is no conflict between the systems themselves which are essentially complementary: the old and the new.

Gumede (1990:207) further notes that the international community as personified by the World Health Organization, through extensive international surveys, has arrived at the following conclusions.

(a) Traditional medicine caters for most of the health needs of at least 80% of the African population (WHO Report 1983). Gumede also notes that over 80% of Black patients visit the traditional healer before going to the hospital or doctor.

(b) Traditional systems of medicine remain the major source of health care for more than two thirds of the world's population (WHO Report 1982). Approximately half the world's population have government departments responsible for traditional medicine. Over two thirds of births in the world are delivered by local or traditional midwives or birth attendants.

Such findings are supported by local research, indicating that, where available, modern medicine is the treatment of choice, with increasing reliance on traditional medicine and combined usage of traditional and modern medicine as the setting becomes more rural (Edwards 1989). In one study it was shown that whereas traditional healers were more supernaturally oriented in their interview methods than were clinical psychologists, both healers and psychologists were in significant agreement with regard to the ranking of alternative diagnostic and treatment methods when interviewing the same group of psychiatric patients. Furthermore the patients rated the healers and psychologists as being more or less equally helpful (Edwards 1986).

The topical issue of harmonization of traditional and modern health systems in the international community and particularly in Africa has been much debated over the last two decades. Najman, Klein and Munro (1982) have distinguished between four main organizational relations between official and traditional health care services.

(a) Monopolistic relations give modern medicine sole legal right to practice.

(b) Tolerant relations exist where modern medicine's exclusivity is limited to specific medical and public health activities. While traditional medicine is not officially recognized, traditional practitioners are free to work and be paid for services in all fields provided they do not claim to be medical doctors.

(c) Parallel relations refer to official recognition of both modern and traditional medicine in equal but separate health care systems.

(d) Integrated relations exist where modern and traditional medicine are merged in medical education and jointly practised within a unique health service as in China.

In 1978, the World Health Organization (WHO) adopted the popular slogan "Health for all by the year 2000". As Gumede (1990:iii) notes, this is a worthy cause but the responsibilities and organizational prospects are tremendous.

The prospects for greater collaboration and or integration of modern and traditional health systems in various African countries have been a much-debated topic. Organizations such as the World Health Organization (WHO) have suggested retraining of traditional practitioners to serve as primary health care workers. In such countries as Nigeria, where the ratio of modern physicians to population of 1:22,000, way below the WHO goal of 1:10,000 for developing countries, integration of the two health systems exist in the form of community therapeutic village treatment programs.

In 1974, the South African Medical and Dental Council made official their rejection of indigenous healers in a Health Act which forbade non-registered healers to practise or perform any act pertaining to the medical profession. Registered healers were also forbidden to work in collaboration with non-registered healers. The unofficial view, which, differed from the official one, was outlined in the *South African Medical Journal*, the official organ of the Medical Association of South Africa, a professional, non-statutory organization. It suggested that medical professions should attempt to understand the healers' system of operation, should accept healers in certain fields and should help them to recognize those illnesses they are unable to cure and learn from traditional medicine (Farrand 1980). Unfortunately, the attitudes reflected in the legislation often resulted in many reputable traditional healers being extremely secretive in the practice of their profession.

It should not be construed that healers are unprotected or uncontrolled by law. Provision was made for their practice in the Code of Zulu Law for example. In the Associated Health Service Professions Act (No. 63 of 1982), the Council of Associated Health Professions, a separate statutory body not affiliated to the Medical and Dental Council, provided for the registration and control of healers.

Over the years, modern mental health medical and paramedical professionals, in particular, have urged for more national collaboration of the two health systems, with the view that traditional healers have an important role to play, particularly in the fields of psychology and psychiatry (Rappaport & Rappaport 1981). Various writers have noted, for example, universal components of psychotherapy shared by both indigenous healers and modern professionals (Cheetham & Griffiths 1982). Shortcomings in the prevailing South African mental health services have been pointed out by various prominent modern South African healers, who have appealed for greater recognition to be given to and for greater use to be made of the skills of indigenous healers in treating persons who may benefit from their services (Holdstock 1979; Wessels 1985; Gumede 1990).

The relatively widespread patient usage of traditional medicine, alone or in combination with modern medicine, and the research findings mentioned above, seem to warrant organized collaboration of modern and traditional healing systems as has happened in neighbouring Swaziland, Zimbabwe and

Bophuthatswana. Administratively, this remains a complex issue within the overall South African political, economic, legal and health situation, with its many communities varying in language, modernization, culture and state-allocated health resources. However, in practice, rapprochement and co-operation (if not integration) are effected in many transcultural psychiatric settings throughout the country and seem a viable legalized likelihood in a new democratically governed South Africa.

Certainly there are good and bad practitioners in both systems, as in any profession or occupation. Certainly both systems have a lot to offer each other and a lot to learn from each other, as has been the case in the past. Day-to-day practical co-operation is especially in the hands of those primary health care workers who form grass-roots liaison with indigenous healers. Many traditional and modern practitioners and their patients have benefited by mutual co-operation established in many areas of South Africa. An integrated holistic approach is currently taught at many South African universities. It is hoped that this process, as well as the urgently needed building of more hospitals and clinics, will continue.

As mentioned earlier, the roles of the different categories of practitioners are becoming increasingly specialized and differentiated, with the umthandazi or faith-healer achieving increasing popularity, particularly in urban areas. It has been noted that patients often regard urban traditional healers as quacks compared to their rural counterparts (Farrand 1984). However, charlatans also exist in rural areas, where some unscrupulous healers have been observed, for example, to utilize public address systems to simulate ancestral voices during patient interviews (Sokhela 1985). Although illegal, the use of killing medicines (umuthi wokubulala) continues (Gumede 1990).

In the meantime, much can be gained by continuing local health education for and amongst both traditional healers and modern health care workers, with the goal of enlightened communication and the recognition of expertise and limitations in both systems by both systems.

Continuing specific and nationally co-ordinated research into traditional and modern medicine is needed in South Africa, along the lines undertaken in neighbouring Swaziland, Zimbabwe and other African countries with a greater need owing to relative lack of modern medical facilities (Green & Makhula 1984; Chavunduko 1978; and Du Toit & Abdalla 1985). Bophuthatswana, for example, had the opportunity (and need) to include all healing practitioners in one health act, thus short-circuiting traditional professional exclusiveness. Nationally co-ordinated research should particularly involve those medical, academic and research institutions which already effect practical rapprochement and co-operation between traditional and modern medicine, and which teach both approaches integratively. The following, in particular, are some of the areas that need to be investigated (Edwards 1989:23):

(a) Research into the effects of modernization, education, economic, socio-cultural and political change on traditional and transitional societies, their related health practices and help-seeking behaviour.

(b) Research with large representative groups of patients, traditional healers and modern practitioners to assess the demand for, and problems and implications of greater integration of traditional and modern medicine. The research would have to cover definition of terms, variations of traditional belief systems, the cultural integration of illness, the ethical and legal implications of traditional and modern healers working together, the training of traditional and modern healers, etc.

(c) Longitudinal follow-up studies on matched groups of patients who have received traditional, modern and combinations of traditional and modern medicine respectively would be valuable.

(d) Phenomenologically oriented outcome research is needed to assess the efficacy of traditional/modern medical approaches not only in healing illness but also in providing therapy toward becoming more fully human and healthy.

(e) Further studies on universal essences of holistic helping and healing as well as the relationship between religious, psychological, medical, social, political, economic and other components of holistic healing as, for example, practised by the African Indigenous Churches.

As 80% of all patients consult traditional healers, it has been suggested that traditional healers be included in the mental health team (Newman and Lauro 1982). One way of doing this is through the professionalization of traditional healers. However as Kottler (1988) has asked the question, "who wants to professionalize who and what exactly is meant by professionalization?"

Nzima (1989) has discussed the pros and cons of professionalization from various perspectives, e.g. legalized sanctions, codes and ethics, admission requirements, registration, training institutions and fees, certificates, professional politics, etc. and noted that professionalization involves essentially a setting of boundaries whereby certain people are defined as in or out. In the South African context it would not be appropriate if the pressure towards professionalization came only from established modern medicine which has often historically been regarded as an institution of the oppressive regime.

Nzima (1989) argues for further harmonization of modern and traditional medicine whatever the historico-political undertones, maintaining that the problems are not caused by any practical healing expertise of each model but in the visible working distance which may occur between traditional and modern medicine.

From a religious point of departure, the African Indigenous Churches have great advantage in that they bridge and integrate modern and traditional religious systems in addition to offering a popular model for health care and a unique form of holistic healing.

Conclusion

Healing concerns change from illness to health. Katz and Wexler (1989) argue that traditional healing systems can also be vehicles for social, political and economic liberation. The concept of the "renewed tradition" refers to the use of respected principles of the past forging new guidelines for action in terms of contemporary needs. Though the renewed tradition stresses spiritual and community transformation, it does not ignore moral, political and economic transformation. Healing is a shared resource, a right for the ill to request and a responsibility for healers to deliver.

It seems appropriate to conclude with remarks made by Conco as quoted by Gumede (1990:231)

> "African traditional medicine
> A resource or a liability
> Can it be abolished? No.
> Can it be successfully replaced? No
> Can a co-existence be negotiated? Yes."

"Finally it is anticipated that modern and traditional healers will be able to achieve rapport and share their skills for the common good of their patients so that health for all will certainly be a reality before the year 3000" (Gumede 1990:iv).

Human and helping relations always occur within a changing context e.g. political negotiation, economic bargaining, sociological solidarity, religious faith healing, psychological psychotherapy with specific forms and combinations of forms being especially needed and relevant for certain times and places.

It was Maslow who maintained that the psychologically healthy person is a political necessity (Edwards 1990:27). While South Africans work their way towards a politically negotiated settlement, the African Indigenous Churches are showing one way in a form of healing that is both traditional and transformational to the psychological and spiritual health of the South African community as a whole.

BIBLIOGRAPHY

Berglund, A I 1976. *Zulu Thought Patterns and Symbolism.* Cape Town: David Philip.

Brammer, L M and Shostrum, E L 1982. *Therapeutic Psychology. Fundamentals of Counselling and Psychotherapy.* Eaglewood Cliffs: Prentice Hall.

Buhrmann, V 1977. Xhosa diviners as psychotherapists. *Psychotherapeia*, 31, 17-20.

Chavunduko, G 1978. *Traditional Healers and the Shona Patient.* Gwello : Mambo Press.

Cheetham, R W S and Griffiths, J A 1978. The Traditional healer/diviner as psychotherapist. *South African Medical Journal*, 62, 957-958.

Conco, W Z 1972. The African Bantu traditional practice of medicine. Some preliminary observations. *Social Science and Medicine*, 6, 283-322.

Corsini, R J 1984. *Current Psychotherapies.* Itasca : F.E. Peacock.

Du Toit, B M and Abdalla, I H 1985. *African Healing Strategies.* New York: Trado-Medic Books.

Edwards, S D, Cheetham, R W S, Majorsi, E & Lasich, A 1982. Zulu culture bound psychiatric syndromes. *South African Journal of Hospital Medicine*, 62, 97-99.

Edwards, S D 1985. *Some Indigenous Views on Illness and Healing in South Africa.* University of Zululand Publication, Series B. No. 49.

Edwards, S D 1986. Traditional and modern medicine in South Africa, a research study. *Social Science and Medicine*, 22, 1273-1276.

Edwards, S D 1987. The Isangoma and Zulu customs. *University of Zululand Journal of Psychology*, 3, 43-48.

Edwards, S D 1988. Psychology as dialogue: didactic implications. *University of Zululand Journal of Psychology*, 4, 1-22.

Edwards, S D 1989. Traditional and modern medicine: some reflective and research considerations. In Oosthuizen, G C, Edwards, S D, Wessels, W H, and Hexham, I (Editors) *Afro-Christian Religion and Healing in Southern Africa.* Lewiston : Edwin Mellen.

Edwards, S D 1990. *Everyday Psychology:* Mtunzini: Spatzennest Publications.

Farrand, D 1980. *An analysis of indigenous healing in sub-urban Johannesburg.* Unpublished Master's thesis. University of the Witwatersrand.

Farrand, D 1984. Is a combined Western and traditional service for black patients desirable? *South African Medical Journal*, 66, 779-780.

Green, E C & Makhulu, E 1984. Traditional healers in Swaziland: towards improved co-operation between traditional and modern health sectors. *Social Science and Medicine*, 18, 1071-1079.

Gumede, M V 1990. *Traditional Healers. A Medical Doctor's Perspective.* Cape Town : Skotaville.

Holdstock, T L 1979. Indigenous healing in South Africa: a neglected potential. *South African Journal of Psychology*, 9, 118-124.

Hutchings, A 1992. *Personal communication.*

Katz, R & Wexler, A 1989. Healing and transformation: lessons from indigenous people (Botswana). In Peltzer, K & Ebigbo, P O (eds). *Clinical Psychology in Africa (South of the Sahara, the Caribbean and Afro-Latin America),* Nigeria : Chuka.

Kottler, A 1988. Professionalization of African healers: apparent problems and constraints. *Psychology in Society*, 11, 2-17.

Kruger, T M D 1974. *Xhosa divining and contemporary psychotherapy: a reciprocal perspective.* Fort Hare Papers, 6, 37-47.

Kruger, D 1984. *The changing Reality of Modern Man: Essays in Honour of Jan Hendrik van den Berg.* Cape Town : Juta.

Lee, S G 1969. Spirit possession among the Zulu. *Spirit Mediumship and Society in Africa*, Beattie, J & Middleton, J (eds). London : Routledge and Kegan Paul.

Mbiti, J S 1970. *African Religion and Philosophy.* London : Longman.

Murdock, G P, Wilson, S F & Frederick, V 1980. World distribution of theories of illness. *Transcultural Psychiatric Research Review*, 17, 37-64.

Najman, J M, Klein, D & Munro, C 1982. Patient characteristics negatively stereotyped by doctors. *Social Science and Medicine*, 16, 1781-1789.

Newman, A K & Lauro, P 1982. Ethnomedicine and biomedicine linking. *Social Science and Medicine*, 16, 1817-1824.

Ngubane, H 1977. *Body and Mind in Zulu Medicine.* London : Academic Press.

Nzima, D R 1989. *The professionalization of traditional healers in South Africa: a case study.* Unpublished Honours Project : University of Zululand.

Oosthuizen, G C 1986. *Religion Alive,* Cape Town : Hodder and Stoughton.

Oosthuizen, G C, Edwards, S D, Wessels, W H & Hexham, I 1989. *Afro-Christian Religion and Healing in Southern Africa.* Lewiston : Edwin Mellen.

Rappaport, H & Rappaport, M 1981. The Integration of Traditional and Scientific Medicine. A proposed model. *American Psychologist*, 36, 774-781.

Rogers, C 1980. *A Way of Being.* Boston: Houghton Mifflin.

Sokhela, N W 1985. *Personal Communication.*

Toffler, A 1980. *The Third Wave.* Pan Books.

Torrey, E F 1972. *The Mind Game: Witchdoctors and Psychiatrists.* New York: Emerson Hall.

Wessels, W H 1985. Culture bound syndromes in South Africa. *Proceedings of the Fourth National South African Congress of Psychology*. Durban.

West, M 1975. *Bishops and Prophets in a Black City. African Independent Churches in Soweto.* Johannesburg: David Philip.

Wolberg, L R 1977. *The Technique of Psychotherapy.* New York : Grune and Stratton.

CHAPTER 9
A ZIONIST CLEANSING CEREMONY: A THERAPEUTIC MODEL FOR A TRAUMA-LADEN SOUTH AFRICA

by

H. Mkhize

Introduction

In the past few years research on therapeutic practices has produced a series of challenges which need to be carefully resolved in a move towards the establishment of acceptable healing models. The Marxist-socialist approach argues that reality is largely determined by the socio-economic and political conditions in which the person finds himself (Hayes 1987). On the contrary, bias in mainstream therapeutic practice is on individualistic conceptions of psychopathology and as a result it tends to favour person-centred models of healing.

Historically, during the colonial era and the apartheid period, Africans lost their cultural values, material resources and their technology. Logically such a loss had an impact upon people's esteem and their behaviours. Africans dissociated themselves from teachings and practices which were African in origin and yet the discriminatory laws prohibited them from being integrated into the "Western" world. This social arrangement had an impoverishing effect upon people's total life. Given this historical background it is essential to search for healing methods which will also counteract the colonisation process.

The case which is being made here is that in today's move towards awareness and sensitivity about the effects of traumatic experiences in people's lives the healing function of African Independent Churches cannot be overlooked. Acknowledgement in this regard should be given to Oosthuizen (cf. Oosthuizen, Edwards, Wessels, Hexham, 1989) who has over years mobilized and in some instances induced interest in researchers with different persuasions to examine the position and function of African Independent Churches in society. It is significant that he committed himself to this task at the time when it was not fashionable to identify with Black people's struggles.

What seems to be a consistent theme in most studies which have looked at African Independent Churches is that their ceremonies provide a therapeutic context within the impoverished social context (Katz and Wexler 1989; Mkhize & Arosi 1990; Mkhize 1991).

In the past few years research on therapeutic models has produced a series of interesting views. Most authors have criticised the relevance of individually based approaches in psychotherapy within a multi-cultural South Africa (Dawes 1985; Hayes 1987). However, many African societies have informal healing practices, like the healing procedures of the Zionist Church, which have been completely ignored in Western-based formal health religious and educational policies. The healing function of African Independent Churches highlights a challenge which still faces even progressive social service organizations. They should guard against the tendency to offer social services as though the African community was an empty 'terrain' with no history, no culture, no knowledge. A similar concern was expressed by Nzimande (1988).

Case Analysis

In this paper I present a description and analysis of a healing ceremony organized by a Zionist Church in Soweto. A well-known leader of the 80s had been in exile for the past ten years. His mother had been a great prophet of the Zionist Church. Both his parents had died while he was in prison. On release he stayed with his younger brother who had been looking after the house since the death of their parents. The returnee was beginning to feel vulnerable. He had been through a difficult time and wanted to talk to other people about some of his experiences. He often saw people passing by but they seemed not to know what to say to him. Silences after each greeting made him feel agitated and he was beginning to anticipate the worst from other people. He found it extremely difficult to sleep. He developed a habit of locking himself in his bedroom to avoid his brother's visitors. His brother became fearful of him and he decided to inform a woman who was friendly to his late mother. The problem was discussed among church members and a decision was taken to have a ceremony for him so that there could be a reconciliation between him and other members of the community and also his late parents. The rationale behind this move was to integrate the returnee within the community at large and to reconcile him with his ancestors.

The Zionist ceremony is basically coercive in tone. The healers are permissive, sensitive and their ability to listen and to reflect on the spoken words is high. The structure of such a religious ceremony resembles that of the traditional ceremony of diviners. Members of the Zionist Church often make use of open spaces for their ceremonies. Hence their healing sessions are public. They attract members as well as non-members.

Singing and dance are essential ingredients during the healing activity. A healing ceremony could be organized for a person with a physical or psychological or social problem. For a Zionist a person is always viewed as a complete whole; levels of symptom manifestation are not significant.

Healing is initiated by religious leaders who get into a highly spiritual state and sometimes even "speak in tongues". They call upon the participants to come to the fore with their problems and they then interpret and guide the person about how best to cope with the problems. The onlookers actively facilitate the healing through music and dance. Implicit in this practice is the belief that complete healing can only take place within a person's total context.

In the healing ceremony under review, the returnee had to stand in the centre and one of the chief healers introduced him and briefly touched upon his problem. He also had an opportunity of talking openly about his problems. He expressed feelings of helplessness and apathy related to a sense of not belonging to the community, a sense of guilt that maybe his parents were disappointed in him and that they did not bless him before they died. As he further talked, there were instances when it had been difficult for him to sleep in fear of his brother, most probably because he suspected that he could do him harm. He also did not trust members of the community, and he feared that he could be attacked at night. At one point he said, "If you have just come back, it is difficult to trust anyone, the incidence of violence is worrying, everything is confusing; there is no one to trust, there is no direction. The heroes who died in exile are better off. At times one gets so frustrated and thinks that it will be better to commit a serious crime so as to go back behind bars. I need policemen to save me from my emotions".

Experienced members of the Zionist Church and many other highly emotional people began to assure him saying; "We love you; your parents were part of us; we know all about their anger and their disappointment in you; they took you like a prodigal son; but now that you are back, we should all celebrate".

For me it became evident that the returnee and members of the community stood at two opposite poles. It became essential for each to understand the other's position. The returnee had to realize that his disappearance was a shock and experienced as a form of rejection by his people. On the other hand, in exile he encountered severe difficulties, most of which were highly unexpected and traumatic. Back home he found himself re-living some of those experiences and he expected people to understand and to be sympathetic.

It became essential for me to help them to see each other's position and particularly to talk about what the returnee was going through. Reference was made to what the *Diagnostic and Statistical Manual of Mental Disorders* of the American Psychiatric Association, as revised in 1987, is referred to as Post-Traumatic Stress Disorders.

Group and community support seem to be essential and appropriate in facilitating the returnee's adjustment process. This form of support stops the process of alienation which often starts when a person escapes from his country of origin. Because of the history of politics in South Africa the returnee would always want to establish the political credibility of the helper. Group counselling might be more fitting since it gives an individual trauma victim with a feeling of a shared responsibility and it prevents "paranoidal" feelings which might easily be evoked in individual counselling.

Although at the present moment there are no in-depth studies which have been undertaken to examine the healing function of African Independent Churches, the fact that they are accessible to members of the community put them in a much more stronger position. They have "special powers" but are still part of the community. They share joys and sorrows of their so-called patients. Although they have faith in God, they still uphold cultural teachings. For them the world is an integrated whole. In Ngugi's (1986) words, one would say they have struggled against imperialism by upholding African values like care, respect and awareness of each other's needs. They have consciously upheld such values even at the time when practices which were African in nature and origin were attacked by the apartheid machinery. On the basis of our limited understanding of their functioning and of what transpired in the above-mentioned case, it is quite evident that the African Independent Churches could provide an effective launching pad into the future. These healers operate in such a way that they offer safety and security through personal exploration which is a vital experience for any person, but even more so for those who have been deeply hurt.

Implicit in the above brief exposition is a belief that there is a need to rediscover African "technology". To reiterate this view reference will be made to Chinweizu, Jemie and Madubuike's (1980) suggestion that,

> "......since an imperialist education has tried to equip us with all manner of absurd views and reactions to our past, we do need to reclaim and rehabilitate our genuine past, to repossess our true and entire history" (p.256).

One step to achieve the proposed goal would be to revive group support systems within trauma-burdened communities. This was Scurfield's (1985) emphasis that "individual" trauma has an impact on the family and as such treatment should include other family members as well. Group treatment has the following advantages:

(a) it reduces isolation by providing a sense of community comfort and support;

(b) it reduces the feelings of stigma and restores self-pride;

(c) confrontation by peers who have suffered similar experiences is often more acceptable and reality-oriented;

(d) it creates an opportunity to work through "unfinished business" and makes shared problems more understandable and hence more acceptable; and

(e) it provides a supportive and understanding environment which facilitates the free expression of emotion. (Scurfield, 1985, p. 247).

Trauma within this intervention strategy becomes a social phenomenon rather than a form of individual "pathology".

BIBLIOGRAPHY

American Psychiatric Association 1987. *Diagnostic and Statistical Manual of Mental Disorders*. Washington, D.C: APA.

Chinweizu, Jemie O and Madubuike, J 1980. *Toward the Decolonization of African Literature.* London: Melbourne & Henley.

Dawes, A 1985. Politics and mental health: The position of clinical psychology in South Africa. *South African Journal of Psychology,* 55.61.

Hayes, G 1987. Challenges to psychotherapy in the context of social service work. *OASSSA National Conference Proceedings*.

Katz, R and Wexler, A 1989. Healing and transformation : Lessons from indigenous people (Botswana). In *Clinical Psychology in Africa* by Peltzer, K and Ebigho, P (eds.) Nigeria. Emugu.

Mkhize, H and Arosi, N 1990. Mental health transformation : Lessons from African Independent Churches. Paper read at NERMIC symposium.

Mkhize, H 1991. Towards the emancipation of women in a post-apartheid S.A. : A case of a prayer woman. Paper read at NERMIC symposium.

Ngugi, Wathongo's 1986. *Decolonising the mind*. London: James Currey.

Nzimande, B 1988. African life and the "hidden abode" of mental health : Some unasked questions about "tradition" and progressive social services in South Africa. *OASSSA Third National Conference Proceedings.*

Oosthuizen, G C; Edwards, S D; Wessels, W H; Hexham, I 1987. *Afro-Christian Religion and Healing in Southern Africa*. African Studies Vol. 8. Lewiston: The Edwin Mellen Press.

Scurfield, R M 1985. Post-trauma stress assessment and treatment: Overview and formulations. In C.R. Figley (ed.) *Trauma and its wake : The study and treatment of post-traumatic stress disorder.* New York: Brunner/Mazel Publishers.

PART THREE

ON RELIGIOUS COMMUNICATION IN AFRO-CHRISTIAN CONTEXT

CHAPTER 10
MUSIC AND LEADERSHIP IN THE ZIONIST CHURCHES

by
M. Xulu

Of all activities in the Zionist Christian worship, music is the most dominant. Few people deny that more than anything else: a Zionist congregation sings. Members of these churches are also very proud of their musical abilities and, in most cases, have high praise for good singing.

According to the Rev Mfeka of the Zion City Church at Umlazi, "Zionist singing can be divided into spiritual singing (ukucula kukamoya) and general congregational singing (ukucula nje okujwayelekile). Spiritual singing is usually done just before *ukuthandazela*, that is, praying for the sick or as part of it". (Interview, 29 October 1991).

Song leaders in spiritual singing are normally *abakhokheli*, that is, the church mothers, and the priests. In other words, it is the most senior people in the congregation who lead in spiritual singing. General congregational singing is mostly led by the young and more physically active *abavangeli* (the evangelists) and *abashumayeli*, the preachers, as well as other distinguished members. Seldom, however, are songs led by new church members and children. For this paper we shall refer to spiritual singing.

Spiritual singing (ukucula kukamoya)

Spiritual singing is generally characterised by a slow tempo, internalised and highly emotional cross-rhythmic patterns, internal improvisation, restrained body movement, open, resonant vocal quality, and high-pitched vocal levels. Often no drumming is involved.

Zionist spiritual singing is the highest form of such Christian worship, and as such, it is surrounded by an atmosphere of dignity. It is the most senior people in the congregation who participate in it and who understand its deeper significance. Thus, they lead in such singing. Children and other more physically active members are at this stage pushed to the background. In some cases they may simply not sing, because of the complexity of the style of singing.

Spiritual singing is marked by yawning, speaking in tongues, prophesying and a general state of transformation. Often it is at the height of spiritual singing when people are sweating and some are even crying, while others may be lying low on the ground, that the holy spirit is felt to be present. At this point those who have opened up for the holy spirit receive enormous powers from God through the priest or bishop and they are capable of doing things they would normally not do because their trance state reveals to them such secrets of the world as can not be revealed in any other way. It is at this stage that personifications of Christ and even God in the human form of priests and archbishops can be witnessed. This level presents the ideal state of Zionist Christian worship. All energies in all other forms of Zionist religious expression are geared towards reaching this climatic point.

It is also at this stage that the sick are healed, the weak are strengthened and deepest prayers of all kinds are said. However, this does not happen in every service and every act of worshipping. It is, however, essential that it happens, for it is through spiritual singing and activities surrounding it that the Zionist idea of man's relationship to man and to Christ and God is realised. In that way the meaningfulness of the gospel is revealed. It is also at this stage that ordinary members in the Zionist Church learn to understanding the nature of Zionism, and identify with it even more.

Zionist congregations, which easily and often transform to reach this level of Christian expression, grow numerically and spiritually, because the ultimate aim of Zionist Christian worship is to understand man's relationship to man and to God and Christ. It is also to solve the daily problems of the church members and make them understand life even better. People identify with congregations which match with these ideals.

Thus, at this level, philosophical manifestations of the nature of the gospel are revealed. It is a level of least verbal arguments and misunderstandings. It is also a level at which the church leaders clearly establish their position of leadership and authority.

A pattern exists which relates musical ability to congregational leadership in the Zionist Churches. Firstly, the nature of the music, which is mostly call-and-response, clearly establishes patterns of leader and followers. What is even more significant is that at the height of singing, those who are leaders in the church become leaders in the music as well. The idea is that a leader leads people somewhere, and in song the person who initiates the song, initiates a path to be followed. He must be a skilful person, somebody the congregation trusts, not only as its spiritual leader but also as a musician, the idea being that if something goes wrong with the singing, the transformation necessary for people to relate to God and to one another meaningfully many not be reached. In Zionist thought the song leader leads his chorus to heaven or in that direction. People "cross the river Jordan", or "reach the Canaan" or even "climb the mountain of heaven" through song. The idea is, therefore, that the song leader, must be the person nearest to God, who knows "the way to heaven". The idea of people triumphantly entering heaven is often enacted in Zionist congregations.

The second point, which relates to the above, is that the philosophical world which gives meaning to the musical patterns of the Zionists is the traditional African world, which establishes socio-religious leadership patterns along the lines which relate to musical ability. Thus, music and leadership in the Zionist Churches has to be interpreted against the traditional African philosophies of music and leadership, which have a profound influence on Zionist through patterns and behaviour. In fact, there is a pattern which links Zionist Christian practice with traditional African thought.

Music and leadership patterns in traditional African society

In most African societies musical ability and social leadership are intertwined. Music here is given the broadest possible definition. The mystic power surrounding music makes musical ability one of the stronger factors in determining leadership preferences.

Let us take the case of a young Zulu child who grows up in a traditional setting as our example. We shall use the masculine to include the feminine. He receives his first music lessons from his grandparents and other people in the community. The lessons are coupled with basic philosophies of life and insights into the nature of the world, which he will further piece together and enrich on his own as he grows. He grows amidst many other children who also receive similar instructions. Soon he understands that society demands him to establish his individual identity. Through asserting his individual identity he has to display his understanding of himself, the people around him, the surrounding environment, God and the ancestors. Although at an early stage some of these factors may be vaguely understood, the child soon learns that without the participation of all these factors in his life, he will live in a world of meaninglessness.

He learns the significance of other people in his life. It is, however, in music-making that this and many other factors will exist and be realised at a concrete level. To take this issue of the role of other people in one's life further,

Zulu music is ordered in such a way that without the participation of others, no individual can meaningfully make music. This means that regardless of the amount of talent one may have, one needs the participation of other people, thus, the call-and-response. The Zulu people have coined the proverb: "*Umuntu ngumuntu ngabantu,*" meaning, "one is human because other human beings participate in his life and he participates in theirs". This proverb largely continues to influence Zulu thought patterns and those of many Africans even in the modern world. It forms the underlying philosophical basis for African music-making, and general African thought. This explains the value for human life and social relations. The growing child is continually being educated in this regard.

In Zulu thought one also has to relate to the environment. Knowledge of birds, trees, seasons and other issues pertaining to nature is hailed as invaluable knowledge. The environment is important in carrying out religious practices. It is also music-making which, more than any other, realises the idea of man and the environment. Some musical events demand direct relation with the environment. Thus, weddings take place in specific seasons and not in others.

Through the singing of *amahubo* songs, the ultimate form of Zulu musical expression, the Zulu idea of man's relationship to man, to God and to the ancestors is realised. Although the Zulu God does not influence life patterns on a daily basis, the ancestors do and their presence is felt at the height of *amahubo* singing.

What is significant is that because of their religious nature, *amahubo* songs are generally sung by the senior members of the clan lineage with the homestead or clan heads as the leaders of the singing. It is because of their nearness to the ancestors that they assume this role. Elders are sometimes regarded as a reflection of the ancestral world.

What is even more significant for a growing Zulu child is that receptivity to music poses further social challenges which iF he meets them, give him a fair chance of being accepted unanimously by the peers as their leader. Music puts people on display and others are able to assess them objectively to see whether they meet the criteria for social and sometimes political leadership. Thus, even today among the Zulu, people who are accepted as socio-political leaders are also great musicians. Inkosi Mangosuthu Buthelezi and the Zulu King bear testimony to this.

A pattern relating to the traditional African world as a basis for the Zionist interpretation of the Christian message can be traced. The fact is that the Zionists are demonstrating that in Africa Christianity has not been introduced on a blank slate. There is then natural syncretism between Christian ideals and the African religious principles. This gives the Zionists their unique identity.

Music and leadership patterns in the Zionist Churches

Although Zionists started as breakers of African tradition, today most of them are extreme traditionalists who draw heavily on their traditional African roots in their interpretation of the Christian message. The closeness of the Old Testament to African thought makes it the most read section of the Bible in the Zionist Churches.

As stated earlier, the very nature of Zionist music invokes, at least subconsciously, the traditional African thought and with it comes all the African thought together with the ideas surrounding music and religious worship. The Zionist priest, bishop or archbishop, because of his image as personification and reflection of God and Christ and, alternatively, because of his felt nearness to God and Christ, has to lead in song, especially in spiritual singing, to transform people to a level where they will feel God and Christ. Seldom does one come across a Zionist church leader who is not a superb singer.

The idea is that music helps the Zionist church leader to achieve his objectives. In song people are able to take their lives and put them in the hands of the leader, because he knows the way to salvation. Music also transforms experience and takes people to a level where they can feel or even see the meaning of their very existence, which is the reason that takes them to church in the first place. To keep them in the church he has to take them to the level where they will experience God and Christ.

Through music the Zionist priest is able to establish and maintain his position of authority and leadership. This is more so because the people he is leading are conscious of the philosophical basis surrounding the African music-making patterns.

Zionists are adamant that music works them into a trance. It transforms their experiences. African thought patterns which are invariably taken to the Zionist Church demand that the leader of the group enhance his leadership qualities through musical ability. In the Zionist Churches most singing is for spiritual empowerment.

Conclusion

We have seen the role of music in the Zionist Christian worship. We have seen also how music influences leadership patterns. Music-making in the Zionist Churches has to be assessed against the traditional African background which has profound influence on Zionist thought patterns.

There is little hope for a Zionist church leader who cannot lead in song as he is unable to lead his people to a level where they will experience physically and emotionally the presence of God and Jesus Christ. He is unable to give his followers spiritual power and distribute it evenly among them. If a Zionist church leader is not a good singer, he often lacks the most viable tool for keeping his congregation together and for making it grow, as well as for making it experience God and Christ through the power of music. He may resort to speeches and other devices which are not able to keep people together for a long time. Musicianship, Zionism and spiritual empowerment are factors which cannot be separated in determining that which constitutes a Zionist congregation.

BIBLIOGRAPHY

Berglund A 1976. *Zulu Thought Patterns and Symbolism.* Claremont: David Philip Publishers (Pty) Ltd.

Blacking J 1969. "The Value of Music in Human Experience" in *Yearbook of the International Folk Music Council,* Volume I, (pp 33-71).

Block M. "Symbols, Song, Dance and Features of Articulation : Is religion an Extreme form of Traditional Authority?" *Archives Europeans de Sociologie* XV: I (pp 55-81).

Dargie D n.d. *Xhosa Zionist Church Music.* Cape Town: CTP Book Printers.

CHAPTER 11

RELIGIOUS COMMUNICATION: PRAYER, SACRIFICE AND DIVINATION

by

S. K. Mfusi

We all have had the experience of failing to communicate, of saying something and being met with a blank, uncompromising look from those to whom we are talking. Our first impulse is usually to repeat what we just said, only a little louder, as if the problem were simply that our well-formulated message was obscure. When an increased volume also fails to get our message through, we may resort to question-begging phrases such as "You know" and many other related gestures. If the intended receiver of our message still does not respond as we have expected him to, we may try mentally searching for ways of paraphrasing what we said.

The above illustration depicts a typical social nature of communication, a way in which individuals transmit intended messages, information or requests to others.

Human beings have evolved many different systems of communication, both verbal and nonverbal. Human communication generally anticipates that the receiver will react to the messages. The most common expectations are that the receiver will alter his behaviour, his plans, his judgment or his opinion, or that he will give other information in return.

Communication among human beings is essentially a social affair. It may also serve to establish or reaffirm social bonds. Messages conveyed in speech do not always have direct relevance to overt meanings of words. When we greet another person by saying "How are you?", for example, we do not usually mean to inquire into his state of health, but merely to acknowledge our social relationship to him.

Religious communication with the supernatural

Religious communication, like human communication, is designed to pass on information - in this case to supernatural receivers.

Before one delves deeply into the important aspects of religious communication, it is important that one looks first into the relationship between religion and the supernatural.

All religion is based on the recognition of a superhuman Reality of which man is somehow conscious and towards which he must in some way orientate his life (de Waal Malefijt 1968). Religion is usually characterized as a cultural and social system, but one criterion that differentiates it from other cultural institutions is the fact that it assumes, and refers to, the existence of powers beyond human power and outside the regular processes of nature (Wallace 1966, Nida 1963). These powers, often conceptualized as beings, are called supernatural or superhuman. Their forms and functions vary, but these higher powers are usually conceived as capable of perception. They are thus believed to be aware of human actions: they "hear" man's words uttered in prayer, notice man's ritual observances, "receive" offerings and sacrifice. Thus, human actions are believed to be potentially capable of influencing the supernatural, of evoking supernatural responses of some sort. Clearly, then, religious systems depict the relationship between man and the Supernatural as a social one.

Culturally held cognitions about the nature of the supernatural are basic to any religious system and significantly affect social and religious patterns. Thus, supernatural entities form part of every religious system, but they appear in many different forms and quantities (de Waal Malefijt 1968). It would not be possible to compile a complete inventory of all supernatural beings in all cultures. A more fruitful approach would be to consider some of the commonly held beliefs about the supernatural, for example, God, the ancestors, and so on. Except among some modern sophisticates, the belief in the existence of supernatural power is universal, but there are wide differences of belief as to the nature of this power.

Many supernatural powers to which the generic term god is applied are believed to be self-created, if their origin and existence is questioned at all. According to de Waal Malefijt, gods are individually known, named, and personified; they are recognizable by their given attributes; ... "their numbers within any given religious system may be large or small" (1968:151). The term high god or supreme god is usually applied to the deity that outranks all others. The Zulu speaking people, for example, refer to their high god as uNkulunkulu (Gumede 1990) while the Xhosa prefer the use of the term uThixo to refer to the same deity.

One belief which seems to be dominant to most Zulu and Xhosa speaking people of South Africa is that their supreme god (uThixo/uNkulunkulu) is too remote and distant to be approached, and for this reason, lesser gods or spirits such as ancestral and holy spirits play more important social roles.

While the term supreme God is usually believed to refer to a supernatural being that created the world, and is for the most part individualized, the term spirits would refer to those supernatural powers that appear collectively. The spirit population is too numerous to be counted, and as a consequence, spirits are not usually named individually but referred to by category, for example, the holy spirits, the ancestral spirits, and so on.

As already pointed out by de Waal Malefijt, spirits are usually below gods in rank (1968). For the Christian religion, for example, this idea would suggest that God and the Holy Spirit coexist in one religious system, but that God is the creator and upholder of the universal order. Further, this idea would suggest that the power of creation is of less social importance than the power to intervene in everyday human affairs, such as the growing of crops, health, and other matters directly concerning human welfare. When spirits are believed to be responsible for common aspects of daily life, they loom larger in man's consciousness, and rituals in their honour are more frequent.

Although spirits appear as multitudes, and often act in concert, individual members of their group may operate on their own as well (Mfusi 1990). Most Xhosa and Zulu indigenous practitioners often single out a few spirits as their guardians and helpers.

The social functions of spirits may overlap with those of the gods. For example, spirits may aid or punish, be favourably or unfavourably inclined toward mankind, demand offerings and sacrifices, and be worshipped and invoked.

Interaction between the dead and the living

The idea that the dead remain members of the community is so common that some authors think it is universal in Southern Africa (Gumede 1990). Although the dead are sometimes feared, ritual activities tend to promote interaction between the living and the dead, rather than keeping them away. I have observed that for most black communities in South Africa funeral rites are acts of reincorporation, readmitting the deceased to the community in their newly achieved status. They surrender one set of rights and duties, and assume another. These beliefs develop into ancestor cults. Ancestor cults then bring the spirit of the deceased back into the community and tend to reaffirm or reinforce the connections between the dead and the survivors. As already pointed out by Gumede (1990) the death of the head of the family is followed by the performance of a ceremonial rite of ukubuyisa (to bring back). Beer is brewed for the occasion, and it is believed that once a beast has been slaughtered, the spirit of the dead person has been brought back. If this duty toward the ancestral spirits is for some reasons neglected, misfortune and sickness will be brought to bear on the family. Most people believe that this turn of events is an indication that the spirits of the departed are issuing a polite reminder that their customary rites are overdue.

Of course, the interaction between the dead and the living varies from place to place, culture to culture, and from one society to the other. Some people, for example, believe that even marriage plans must be presented for the approval of the ancestors, and their blessing must be invoked when a journey or any important transaction is being undertaken. Kuper (1963) reports that among the Swazi, each family propitiates its own ancestors at births, or when building a new house.

Addressing the supernatural

In almost every culture in the world there are customary ways of addressing supernatural beings. Such address is generally conducted by speaking aloud while the body is held in a conventional posture (such as folding one's arms, kneeling down, and so on). While standardized gestures intended to express fear, love, respect, threat, or other emotion, the style of speech is also apt to be stereotyped. The nature, purpose, and content of the address also vary with culture, within the culture, with the occasion, the society, and the individual. Such address may be made in public or in private, at a sacred location and with special apparatus such as incense or smoke from a fire to expedite the message. Most important in this section is that the categories of religious behaviour used to address the supernatural are not intended to represent the only possible classification. Indeed, since the choice of level of abstraction must be made intuitively, no one list can claim absolute validity. The following categories of behaviour used to address the supernatural are based, for the most part, on the author's observations, as well as utterances of a number of informants. It is hoped that they will serve as a rough-and-ready classification of the types of behaviours that are observed as religious in most parts of South Africa. They are prayer, linguistic aspects of communicating with the supernatural, divination and sacrifice.

Prayer is perhaps recognized as the major form of verbal communication with the supernatural and appears to be a nearly universal feature of religion. Perhaps one can say that prayer is the type of address that is most readily recognized by the Christian as fundamental. In prayer the supernatural is either asked for a favour or is thanked for a blessing. Prayers may be offered either on behalf of the person praying or on behalf of some other person or persons. I can remember explicitly during my undergraduate years when my uncle who was a priest would pray on my behalf, asking God to please look after me on my journey from home to the university.

In prayer people reveal their attitudes to the supernatural, and thus their culturally learned ways of dealing with these higher powers. Some of the major types of prayer include petition, adoration, atonement, and gratitude.

In petitionary prayers one asks for something, for example, rain, health, peace, success in obtaining a university degree, and a host of other blessings. In some schools or universities, for instance, when the year approaches the semester or end of the year examinations, students gather in halls or chapels to pray for success in their studies. Even the most rebellious and notorious ones would not miss such a religious service. During the times of drought, shopowners and other major businesses in Transkei are required to temporarily close their outlets to enable their employees to attend prayer for rain. These are some of the examples of petitionary prayers.

Adoration or worship deals with the supernatural by acknowledging their higher powers. In a mood of adoration, one pays homage to the gods, praising and honouring them, and expressing devotion and dependence.

In atonement, one prays for the purpose of seeking amends or reparation for injury or wrong. In other words, one seeks reconciliation with God after transgressing his law, and with one's fellow-person after quarrelling (Reader's Digest 1984).

One also expresses gratitude to the supernatural being for kindness shown in particular events, for example, having our daily bread, surviving an accident, and so on.

Prayers can also be individual and communal. Although some prayers can be performed by individuals in solitude (for example, some people may fast in solitude for a number of days), religious behaviour is not an individual matter. A group of people, such as a congregation, meets on some occasions for the joint performance of religious behaviour. Further, the presence of the congregation during prayers or worships may serve as an inspiration to individual members. It has been observed that in most services of the independent churches some adherents may experience dramatic alterations of mental and emotional states. Experiences of vision, faith-healing, and other divine revelations are sometimes attributed to the inspiration which comes from the congregation as a whole (Mfusi 1984). In such instances, music and dancing are believed to play a crucial role in facilitating communication with the supernatural. When the author visited one of the independent churches in Transkei (the St. John's Apostolic Church), the whole service was characterized by singing, clapping of hands, dancing and chanting. It became apparent that once a service begins in such churches, music becomes the vehicle of prayer if not in itself the substance of communication.

Linguistic aspects of religious behaviour, verbal communication in particular, concern the language that is used when communicating with the supernatural. The differences between sacred and secular language range from minor modification of voice to the use of a completely different language which cannot be understood without special training. In most independent churches which the author has visited, the person who conducted prayer would verbally express him/herself politely for two or three minutes. Then, suddenly, the person's tone of voice would change: the pitch rises and the voice trembles. Perhaps these changes in the person's voice, pitch, and so on signify the urgency and importance of the message that is being communicated to the supernatural at that point in time. Mrs N an African independent church leader from the Umqanduli district in the Transkei, explained the rationale behind the change in voice :

> " It is difficult to explain some of these things. Once I start praying, I know that I am communicating with far-away people, and for the most part we are making requests and seeking answers to a number of things. For instance, if I want to ask a question: ndenze ntoni Thixo Bawo (What have I done, Lord, Father?), I must say it loud and clear so that the Lord can hear and understand explicitly what I am asking him for. If I whisper, God may not hear what I say ".

Most people, however, believe that some dramatic changes that take place during church services are more spontaneous than being deliberate attempts to persuade the supernatural to attend to one's requests. Sometimes during healing sessions, while the congregation sings and dances, the priest or prophet engages in intense prayer.

Incoherent mumbling, groaning, sobbing, sneezing, characteristic of some izangoma (traditional diviners), may announce the beginning of a trance, but these sounds do not signify communication with the supernatural. They merely indicate that the preliminaries are taking place. But once the communication takes place, different language may be heard. When communicating with the supernatural, for example, izangoma zabalozi (ventriloquists) are believed to use a language that is not known to the audience. Amakhosi (as the ancestral spirits are sometimes known) speak in a whistling way from the roof of the hut. In such cases, an interpretation of the message content is necessary.

When prayers are highly formalized, and letter-perfect recitations are demanded, ritual speech patterns are less subject to linguistic change than common language. Examples which immediately come to mind include such prayers as those observed from religious services of the Roman Catholic and Anglican churches whereby the whole congregation join after the priest in reciting prayer.

Divination

Divination is probably being practised in every society, even in modern, thoroughly secularized urban areas. Divination is performed when a decision must be made that the actor feels should be based on more information than is available or on more valid principles of judgment than he or she commands. In such a situation, divination is intended to provide the missing information or principle of judgment by direct, if apparently arbitrary, advice from the supernatural (Mfusi 1984). Even the most militant atheist is apt to turn to simple divinatory practices when neurotic uncertainties or real information gaps make choice difficult. Such procedures as flipping a coin to decide between two alternatives - for example, which football team to kick the ball in which direction, which movie to attend, who shall pay the bill, and so on - are the very prototype of divination. Of course, the secular coin flipper regards the rite as merely a means for random selection between equally attractive alternatives in order to hasten decision or obviate dispute. However, to a diviner, such a ritual is believed to reveal something about nature, including its future, which is hidden from the senses and which requires supernatural complicity to unravel.

Divination is defined by de Waal Malefijt as "a form of religious communication in which supernatural powers give or are coerced into giving, direct information" (1966:215). In nondivinatory prayer (see above), subsequent events are generally interpreted as supernatural answers or reactions, but these reactions may be delayed or postponed. Moreover, divination differs significantly from other forms of religious communication in that it does not generally attempt to alter the behaviour or judgment of the supernatural powers, but merely to discover the will and intention of the gods, and thereby to foretell the future or explain past events.

Once supernatural plans and intentions have been discovered, divination may well be followed by prayer or sacrifice in an attempt to alter them.

For indigenous practitioners, including diviners (izangoma - Zulu, amagqirha - Xhosa), abathandazi (prophet/healers), and izinyanga (traditional doctors), divinatory practices play a significant role in diagnosing illnesses (Farrand 1980; Mfusi 1984; Edwards 1987), in deciding the guilt or innocence of an individual, in finding best places for trading, building a new house or in collaborating important social decisions such as a marriage choice or the choice of friends in general. In other words, divination operates in those areas of life where decisions are not lightly taken.

As a form of communication, there are many types or techniques of divination that have been observed among the Nguni (Xhosa and Zulu) indigenous practitioners, and these may vary from one category of practitioner to the other. One type of divination which has been observed is based on the idea that the supernatural powers initiate the communication process and are the senders of messages. We all have observed or heard how the supernatural powers, such as the ancestors, take the initiative to call upon those whom they want to be healers. During her initial call, an isangoma, for example, usually suffers from a series of illnesses which are later diagnosed by a qualified diviner as a call from the ancestors (Berglund 1976; Mfusi 1984; Farrand 1980; Ngubane 1976; Sokhela 1985; Edwards 1987).

In another type of divination, one takes the initiative by asking specific questions, for exampe, when indigenous practitioners are consulted by their clients to clarify or explain some of the issues mentioned above. In such instances, the practitioner has to perform some kind of experiment (for example, throwing of divining bones, depending on the category of indigenous practitioner), the outcome of which is interpreted as the supernatural answer. The practitioner would, for example, take the bones, address and sing the praises of his or her ancestral spirits and throws all the bones down. It is believed that the ancestral spirits would talk to the practitioner through the way the bones lie. The spirits tell him the facts of the case and give him the diagnosis.

The other type involves the idea of verbal communication; a person speaks directly to the gods in a face to face relationship, and the gods answer in words and sentences (Tart 1969; Drury 1982; Rouglet 1985; Elaide 1964). This usually takes place in trance or in dreams. In such cases, the spirits may be believed to take the initiative and speak. One practitioner said that whenever she was unclear about something concerning some of her patients, the spirit would always visit her in her dreams and explain in detail what medicines were required for specific illnesses.

Sacrifice

Sacrifice is considered a form of nonverbal communication with the supernatural. It is one of the most important events that take place in a ritual setting. It occurs in one form or another in most religious systems and, according to de Waal Malefijt, it has received much more attention in the literature than prayer has (1968).

Although the words sacrifice and offering are sometimes used synonymously (for example, Lamla 1971), the former usually involves the slaughter of animals, and even human beings. (It has been reported by the media that some people murder other human beings in the worship of the devil). The latter normally involves the presentation of food or other material objects for consumption by the supernatural. However, the appearance of sacrificial objects is of less importance to the understanding of the nature of sacrifice than the intentions and motivations behind them.

The above-mentioned communication model used for prayer is applicable to sacrifice. Here too, a "sender" and a "receiver" must be present, but as de Waal Malefijt points out, instead of words the sacrificial objects or items are now the symbolic medium of communication (1968). Some informants, however, maintained that sacrifice can be accompanied by some kind of prayer directed to the deity to whom the sacrifice is made. Nonetheless, the meaning of the sacrifice and the expected reactions depend upon cultural symbolism and the given circumstances of the ritual. Ritual sacrifice, therefore, is a form of exchange between man and the supernatural and is in this respect, similar to verbal communication "... the human actor presents goods; the divine receiver reacts" (de Waal Malefijt 1968:209).

Max Weber (1964:26) believed that sacrifice had its origin in "primitive magic", designed to coerce the gods. According to him, the central theme of sacrifice is "do ut des" (I give that you may give). "Whether sacrifice is offered as barter, bribe, homage, or tribute, man expects some results - material or immaterial, direct or indirect, in this world or hereafter" (ibid:27). He also saw the great social importance of sacrifice, the effecting of a *communio* among the sacrificers. To Weber then, sacrifice is always rational, integrative and purposeful.

Lamla has done an intensive study on sacrifices among the Southern Nguni. In a view partly congruent with Weber's, he has defined sacrifice as "the offerings made to supernatural agencies for their consumption, and is normally characterized by consecratory rites and its primary purpose is to gain certain ends" (1971:1).

Another view suggests that sacrifice is characteristically a communal meal in which human participants and the deity share the same food, thereby strengthening the bonds of unity, and that the meal consists of the animal identified in some way with the god himself (Fisher 1964; cited in Lamla 1971).

What is basically suggested by these views is that sacrifice is a means of ritually compelling the supernatural to perform some desired acts. It may also serve as a means of thanking a god for some past favours, and of invoking his goodwill for desired ends. For example, it has become increasingly common that when one purchases a new car or house, some kind of sacrifice (e.g., slaughter of a sheep, goat, etc.) is made toward a supernatural being. According to most informants, this type of exercise is usually aimed at thanking the gods (for most people: the ancestors) for having enabled them to achieve such goals. The sacrifice is also aimed at asking the gods for further protection in future. Most people are convinced that failure to conform to such beliefs may result in a series of misfortunes, including road accidents (in the case of one having purchased a motor vehicle), or numerous burglaries, in the case where one has purchased a new house. This idea of sacrifice is similar to the one observed at the birth of a child. Most Zulu and Xhosa people, for example, believe that when a child is born, a beast must be slaughtered to thank abaphansi or izinyanya (ancestral spirits) that the mother did not die at childbirth and to show gratitude for the child which the ancestral spirits have given to the family. The sacrificial beast is called imbeleko (from ukubeleka: to carry a child on the back), because the animal's skin is used after the sacrifice for carrying the child on the mother's back (Hammond-Tooke 1962; cited in Lamla 1971). Another significance of this sacrifice is that the child will be protected from serious illnesses by the ancestors.

Sacrifice may also serve to pacify or ease relations with the supernatural. Sundkler, for example, cites a case in which two Zulu women were without children. Then the local prophet was visited by an angel who said that the reason why the two women were without children, was because they had neglected their ancestral spirits, and the latter were angry. The angel demanded that they slaughter a bull. After the performance of this ritual, was reported that the two women were then blessed with children (1961:250).

The above examples show that sacrifice is obviously multi-functional and its essence cannot be reduced to one end only. Among the Xhosa, Zulu, and probably other black South Africans, there are many occasions or situations in which sacrificial rites are performed, and it will not be possible to exhaust them all here. What is of importance, however, is that although sacrificial rites have assumed a multiplicity of forms and intentions, their most fundamental purpose appears to be that of effecting a necessary and efficacious relationship for man with the sacred order (James 1961).

And given that sacrificial rites have been practised throughout the history of religion (de Waal Malefijt 1968), there seems to be reason to believe that, as a manifestation of people's experience of the sacred, sacrifice is rooted in the religious consciousness.

Religious communication in states of possession

There is little doubt that the aspects of religious communication that have been addressed are, for the most part, associated with incidents of spirit possession. Diviners and traditional doctors, for example, are believed to be possessed by the spirits of their ancestors, whereas most prophets/faith-healers are possessed by the Holy Spirit. Prayer, divination and sacrificial rituals are all believed to be the cornerstone of indigenous healing practices in many parts of the world, although the way in which they are used may differ from place to place, and from one category of indigenous practitioner to the other. It has been observed, for example, that prayer is used exclusively as a means of religious communication by Zulu and Xhosa prophets/healers (Mfusi 1990), whereas divination and sacrificial ritual are believed to play a crucial and even an exclusive role in communication between the supernatural and the other categories of indigenous practitioners, namely, the isangoma/igqirha and inyanga. However, a practitioner who operates as a prophet/healer, diviner, and traditional doctor in one is believed to use all the three religious communication rituals (Mfusi 1990). A brief explanation of how this takes place within each category of practitioner is warranted here.

Indigenous practitioners are believed from the very onset to be directed by the spirits of their ancestors. While the ancestors are important for all black people, they are particularly vital for indigenous practitioners (Farrand 1980). Indeed, without the ancestors there could be no practitioners, for it is the ancestors who call the healers to their professions, who enter them when they divine, and guide them through their dreams when they are asleep. For such a relationship to be effective, it must be facilitated by the performance of a number of sacrifices. When the spirits approach them to become practitioners, izangoma, for example, usually become very ill. They have to undergo a period of apprenticeship during which many sacrifices are observed.

Although spirit possession has been extensively investigated (Farrand 1980; Mfusi 1984), there seems to be little agreement to what it actually involves. Much confusion has arisen from the difficulty to categorize the possession states, for the same state is often known by different names and there is also much variation in the behaviour, customs and beliefs about possession amongst different ethnic groups of South Africa (Farrand 1980). However, there is little doubt that izangoma and izinyanga are possessed by the spirits of their ancestors, whereas most prophets/ healers are believed to be possessed by the Holy Spirit. The idea of the Holy Spirit is based on the view of the Christian tradition: the view that a spirit of God may take over some aspects of the individual's functioning. This relatively recent view has found its most popular expression in the African Independent Church movements. Such themes have not been reflected in the most popular literature, probably because of a good deal of strong popular belief that exists. Most independent churches believe that speaking in tongues, for example, reveals the work of the Holy Spirit in the person over whom such individual has only limited control.

Holy Spirit possession is widely found during church services of a Zionist or Pentecostal type. Faith-healing in many independent churches takes place during services, and it is not necessary for the healer to know of any specific complaints of individual patients. Patients are healed through the power of the Holy Spirit which acts through the agency of the healers, and often of the church congregation as a whole (West 1975). Possession in this category is not as continuous as in other forms of possession, for example, ancestral spirit possession. However, as further pointed out by Sibisi, a prophet or faith-healer in such a sect may be thought of as being in continuous contact with the Spirit, in which case his or her role is seen as more or less identical with that of an isangoma (1975).

The way in which spirit possession operates within indigenous practitioners has been observed by other researchers as well (Oosthuizen 1983; Farrand 1980). Some practitioners report that their spirits help them to feel in their bodies what troubles their patients. For example, if the patient suffers from a headache, a practitioner would feel the same ailment in his own body. One prophet/healer from Kwa-Dlangezwa near Empangeni, explained how the spirit helps him in diagnosing his patients' illnesses:

> " ... when people come to me, I ask them to bring a bottle full of water. When I look at the bottle, I see something like a television. I see all the problems on that screen".

Of further importance in the way indigenous practitioners communicate with their spirits, is the use of dreams by the practitioners. In addition to the significance of dreams in the personal lives of the practitioners, dreams are also assumed to play an outstanding role in the diagnosis and treatment activities of such practitioners. Many diviners report that it would be impossible for them to work without dreams. As Berglund points out, dreams are believed to be their "eyes in the work" (1976:98). It is through dreams that the practitioners receive instructions from their spirits. In the diagnosis and treatment of disease a practitioner, for example, would dream of an ancestral member who would show him or her the types of medicines that were to be used in order to effect a patient's cure. Oosthuizen (1988) also observes that in some instances a prophet may have a dream in which the ancestors or "the spirits" reveal the patient's illness before the latter arrives. In short, the practitioners act as mediators for the spirits, the accent being on communication, and dreams are believed to play a crucial role in such cases.

Concluding Remarks

Religious communication has been treated here as an attempt of human beings to convey messages to the supernatural. These attempts have important social implications. Prayer, ritual, and divination unite the participants by the expression of common concerns. As such religious communication contributes greatly to the internal cohesion of the group. Prayer and sacrifice generally involve concern to secure a favour of the supernatural beings for the major enterprises of the community. The inauguration or completion of any enterprise is nowadays apt to be celebrated by a humble sacrifice or offering of something of value to any deity who has a measure of control over that area of experience. Thus, the purchasing of a new house, the celebration of a marriage, opening of a store, and so on, are sometimes believed to require giving up of some valued thing.

Divination is of special importance because it is carried out in situations where decisions or plans of action are not lightly taken. Indecision is thus eliminated, and the tasks at hand can be carried out without delay and further distraction. But more than that, divination stamps the decision thus taken with a special mark of legitimacy. In doing so, it establishes an effective consensus upon the project to be undertaken, be it the determination of a site, the choice of a friend, and so on. As Park (1963:95) also points out, divination has, as its regular consequence, " the elimination of important sources of social disorder".

Finally, and perhaps of vital importance, is the role that is played by the three aspects of religious communication in the African Independent Church movements. Prayer, ritual, and divination play a crucial role in the activities of such churches and are prevalent in most parts of Southern Africa, and have been commented upon by many authors (e.g. Edwards 1983; Mfusi 1984; Oosthuizen 1983). For example, most independent church followers report that they have joined these churches for the purpose of being healed; and the three aspects in question are crucial for almost all healing rites in the independent churches. In one instance which the author witnessed, a man who believed he had a string of misfortunes, such as losing jobs, unpopular with females, and so on, consulted an umthandazi to divine for him. After divination, the man was instructed to bring a live chicken which the umthandazi prayed for. The animal was later slaughtered and buried on the ground (presumably burying his bad luck). In addition, the man was given holy water which he had to drink and wash with every morning and night.

Even more crucial is the fact that most faith-healers/prophets had themselves been victims of long periods of illnesses which had to be cured by other qualified practitioners, and they had to perform a series of rituals for this purpose. Further, many abathandazi are former diviners who had converted to Christianity, and to independent churches in particular, because they had no desire to wear the "skin" attire that is characteristic of izangoma.

Although a large amount of work has been done with respect to some aspects of religious communication, there seems to be a clear research mandate to further investigate some of these aspects in more detail.

BIBLIOGRAPHY

Berglund, A I 1976. *Zulu Ideas and Symbolism*. London: C. Hurst & Company.

de Waal Malefijt, A 1968. *Religion and Culture. An Introduction to Anthropology of Religion*. New York: The MacMillan Company.

Drury, N 1982. *The Shaman and the Magician: Journey Between the Worlds*. London: Routledge & Kegan Paul.

Edwards, S D 1987. The Isangoma and Zulu Customs. *Journal of Psychology*. University of Zululand, 3, 43-48.

Elaide, M 1964. *Shamanism: Archaic Techniques of Ecstasy*. Princeton: Princeton University Press.

Farrand, D 1980. An Analysis of Indigenous Healing in Suburban Johannesburg. Unpublished Masters Dissertation, University of Witwatersrand.

Gumede, M V 1990. *Traditional Healers: A Medical Doctor's Perspective*. Johannesburg: Skotaville Publishers.

Hammond-Tooke, W D 1962. *Baccha Society*. Cape Town: Oxford University Press.

James, O E 1961. In Lamla, M 1971. *Fort Hare Papers*. Alice: University of Fort Hare.

Kuper, H 1963. *The Swazi, A South African Kingdom*. New York: Holt Rinehart and Winston.

Lamla, M 1971. Ancestor Worship: Sacrifices and Offerings Among the Southern Nguni. *Fort Hare Papers*. Alice: University of Fort Hare.

Mfusi, S K 1984. The Role of Dreams for Zulu Indigenous Practitioners. Unpublished Masters Dissertation, University of Zululand.

Mfusi, S K 1990. Spirits, Dreams, and Indigenous Practitioners: An Essential Link. Paper presented at Nermic Symposium, University of Witwatersrand.

Ngubane, H 1976. *Body and Mind in Zulu Medicine*. London: Academic Press.

Nida, E A 1963. *Customs, Culture, and Christianity*. London: The Tyndale Press.

Oosthuizen, G C 1988. Indigenous Healing within the Context of Independent Churches. *Afro-Christian Religion and Healing in Southern Africa.* Lewiston: The Edwin Mellen Press.

Park, G K 1963. Divination and its Social Contexts. *Journal of the Royal Anthropological Institute of Great Britain and Ireland*, 93, 195-209.

Reader's Digest. 1984. *Great Illustrated Dictionary*. London: Reader's Digest Association.

Rouglet, G 1985. *Music and Trance: A Theory of the Relation between Music and Trance*. Chicago: University of Chicago Press.

Sibisi, H 1975. Spirit Possession in Zulu Cosmology. In West, M and Whisson, M G (Eds.). *Religion and Social Change in Southern Africa*. Cape Town: David Philip.

Sokhela, N and Edwards, S D 1985. Natural and Supernatural Healing by Zulu Indigenous Practitioners. *Journal of Psychology*, Series B No. 49. Kwa-Dlangezwa: University of Zululand.

Sundkler, B G M 1961. *Bantu Prophets in South Africa*. London: Oxford University Press.

Tart, C T 1969. *Altered states of Consciousness: A Book of Readings*. New York: John Wiley & Sons, Inc.

Wallace, A F C 1966. *Religion: An Anthropological View*. New York: Random House.

Weber, M 1964. In de Waal Malefijt, A 1968. *Religion and Culture*. London: Collier Macmillan Limited.

West, M 1975 *Bishops and Prophets in a Black City: African Independent Churches in Soweto*. Cape Town: David Philip.

CHAPTER 12
THE INTERFACE BETWEEN BIBLICAL REVELATION AND EXTRA-BIBLICAL REVELATION: AN AXIOM IN THE DEVELOPMENT OF AFRICAN INDIGENOUS CHURCHES

by
S. W. D. Dube

Introduction

Our definition of the new religious movements as an area of study was effectively developed by Turner (1967, 1971, 1974, 1977, 1979). Obvious in Turner's pioneering work is the emphasis that this phenomenon has universal dimensions. Whilst Turner's definition was that of a scholar who viewed the phenomenon from outside, Makhubu (1988) adds an important aspect to the definition. His position is that of a practitioner who defines something that he understands from personal experience. The upshot of this approach is the inevitability of juxtaposing the Bible to the spiritual awareness of the African.

This essay is exploratory. It is intended, in general terms, to encourage ongoing research into the contribution we can make to the history of religion by systematically studying the various responses of the African people to the Word of God which they encounter primarily in the Holy Bible. The specific purpose of the essay is to enquire into the interaction between Biblical revelation and extra-Biblical revelation with special reference to the African Independent Churches.

Several reasons account for the limited focus of this essay. The time constraints as well as the request that the essay be kept brief have led to our concentration on literature which specifically deals with the African phenomenon. It is hoped that this will be seen as an introduction to a field of study which promises to be interesting and exciting.

An attempt is made in this essay to revisit the concept of revelation. This is prompted by the realization that Biblical revelation is enshrined in the written word which is not readily accessible to the membership of the indigenous churches. The latter are believed to be predominantly an oral community, yet their contact with the written Word of God has triggered off a spiritual response which has changed the course of the history of religion. The essay also pays attention to both the historical context which either prompts or accompanies the envisaged spiritual response, as well as the extent to which this spiritual response makes an impact on the spiritual status of the African communities.

Revelation

The Oxford English Dictionary defines revelation as 'the disclosing of knowledge to man by divine or supernatural agency'. Obvious in this definition is divine or supernatural intervention and its role in the acquisition of knowledge. Implied in this definition is the notion that such knowledge is either "secret" or "hidden" knowledge - the kind of knowledge which is not readily available to everybody.

(a) Biblical revelation

Much work has been done on the subject of Biblical revelation. The emphasis in most cases is biased towards a Christian approach to the whole issue. Karl Rahner observes that revelation has divine origin, and as such it is the innermost core of human history (1966:11). Rahner's argument revolves around the fact that in the relationship between God and man God is immanent and this makes him to be superior to the world. It is the quality of God to confer on human beings an active self-transcendence in the change and becoming (Rahner 1966:12).

Rahner's thesis puts emphasis on the fact that revelation results from the self-unfolding character of God. This is prompted by an interior elevating grace which hinges on faith. Consequently two forces are at play in revelation, viz man's supernaturally elevated transcendence and man's grace-borne destiny (Rahner 1966:13,14). At the same time, revelation is occasioned by God's will, i.e. it is a historical mediation which is an expression of the supernatural and transcendental experience (Rahner 1966:14).

The idea of revelation in contemporary thought has been discussed at length by several writers. Among these John Baillie (1956) seems to be pertinent to the discussion raised in this essay. For Baillie, revelation literally means an unveiling, the lifting of an obscuring veil, so as to disclose something that was formerly hidden (1956:19). While this would be true of all revelation, Baillie makes a distinction which qualifies Biblical revelation. In the Bible, revelation implies the disclosure of what has been kept secret (a mystery) for ages (Baillie 1956:28). At the same time Biblical revelation is concerned with the nature and mind and purpose of God. This does not consist in mere information but its content is God incarnate who is revealed in Jesus Christ and who is offering himself in fellowship (Baillie 1956:28).

Our idea of Biblical revelation is traced from the Old Testament. Smith (1936:92) suggests that the Israelites knew God as Elohim who had a name i.e. Yahweh. Smith's emphasis on the name stems from the fact that he contends that a name does not only mark off one person from another, but it is part of that person. It accords that person an appropriate status. Consequently when

God revealed himself to Moses and revealed His name to Moses, a new religion was started. This new religion and the accompanying ways of worship were commensurate with the new revelation of God (Smith 1936:96). Our understanding of this development is in line with the idea of God communicating himself in absolute, merciful presence as God, the absolute mystery (Rahner & Ratzinger 1985:13).

Neuner (1967:39,41) reminds us of the role of priesthood in the development of the concept of revelation. It was through the priestly tradition that God "granted access to himself". This was not just limited to Israel, but it obtained outside Israel as well. This is made possible by the fact that God's grace is active and at work even outside Israel.

The tangible evidence of revelation is found in Biblical prophecy. Shorter (1983:60) suggests that God used persons of certain charisma to speak his word in their voice. In order for this to be possible God penetrated their lives. These persons, therefore, helped Israel through messages and leadership to decipher the meaning of the events that had occurred in the life of the nation. The upshot of the messages and the media through which they were conveyed was that the "otherness" of Yahweh was clearly revealed. At the same time Yahweh's plan for the nation was revealed.

The above brief exposition on the ideas on Biblical revelation points to the fact that God "speaks" and that it is the nature of the spoken language to communicate a message and also to reveal the personality of the speaker (Shorter 1983:40). Our observation has been that revelation takes place within and through a human experience (Shorter 1983:41). Consequently the process involves reflection, understanding, concern and a decision regarding the unveiling of something hidden (Shorter 1983:41). The few persons who become instruments of revelation develop a special relationship with the divine and that relationship is a consistent response to the revelation. What this means is that

Biblical revelation defines a personal communication between God and man and, in the case of Israel, this suggested the idea of election (Shorter 1983:41,42).

Shorter (1983:147,148,212), quoting Karl Rahner suggests that the foundation of the church can be traced to the inspiration and special divine activity where the Bible did not play a minor role. The emergent church, raised on the life of faith, was nurtured in a context of growth and becoming, which pointed to the continuation of the revelation of God. Shorter (1983:212), quoting Taylor, argues that in the context of the life of faith we have evidence of the intervention of the Spirit, i.e. the "beyondness" who opens up vistas for knowledge and increased relationships. It is through this intervention that our personality and our discernment are activated, ushering in possibilities of a fuller participation in the revelation experience.

The foregoing is but a brief attempt to illustrate that the Bible is evidence that one is not alone on the long journey through history. In spite of all that one does to the truth, viz disfiguring it and perverting it, the fact remains that God's revelation continues (Neuner 1967:3,14). The church as an institution is ordered upon the basis of the New Testament message (Neuner 1967:35). The spread of the gospel and the missionary work of the church have opened up the church so that in its interaction with the world it has elicited responses which are found particularly in new religious movements.

2. Extra-biblical revelation

Walls (1986:182) suggests that the force of gravity of Christianity is shifting to Africa in the foreseeable future. In the vanguard of this development are the new religious movements. In spite of the commonly accepted notion that African adherents of new religious movements are relatively illiterate, their faith and their worship in this new religion are characterized by a marked degree of biblicism.

Barrett (1968) discusses the factors that contribute to this development. Chief among these is the translation of the Bible in vernacular.

The availability of the Bible in vernacular did not cancel the position and the leadership which emerged in response to the Good News. Co-incidentally there emerged in the African new religious movements an inspired leadership, charismatic in nature, which did not only reflect on the Word of God as it impinged on the real lives of the people, but also as a prophet the leader became a mouthpiece and a "nurse" and a "midwife" for the birth and the development of a new religion (Barrett 1968). The point I am making here is that while the Bible became important and its presence became exalted even among those who could not read, the position of the spiritual leaders became equally important. It is argued, for instance, that the leader of the Cherubim and Seraphim Church in Nigeria was illiterate and so was his secretary, yet the leader and prophet quoted with ease from the Bible (cf Turner 1967 Chapter 12).

Martin (1964) provides us with some insights into the role of extra-biblical revelation in the development of African indigenous movements, particularly in Southern Africa. Of particular relevance to this essay is Martin's definition of the prophetic messianic churches. This definition is underpinned by three considerations: the emergence of a prophet who becomes in the course of time a saviour, a promise of salvation which revolves around the prophet's activities, and the nurturing of eschatological ideas of the coming bliss and the transformation of all the things which are important in the lives of the faithful (Martin 1964:103).

A number of examples are used by Martin to highlight the aspect of extra-biblical revelation. In the examples used it is obvious that the Word of God is directly accessible to the inquirer. Dreams, visions, ecstasy, etc become important contexts for divine communication. Briefly stated, Biblical and Christian material provide a framework for an interpretation of the interaction between people and their environment, particularly where their innermost attitude is affected (cf Neuner 1967:13).

The examples used by Martin (1964) and Lea (1925) give evidence of the outcome of missionary activity. I would not go as far as Martin to suggest that this also is a preparation for missionary activity (1964:105). To argue in this fashion would be to fall into the same trap that Catholic theology fell into when it was confronted with the emergent Protestantism.

The thesis of this essay hinges on the acceptance that revelation as such is the historical self-unfolding of "the divine" or God in premeditated terms. In this sense it explains the history of the transcendental relation between human beings and God which is enshrined in God's self-communication (Rahner & Ratzinger 1966:13). The limits of the accessibility of this communication are outside the scope of this essay.

In our quest for African theologies we identify the indigenous churches as a valuable source for the development of African Christian theologies. This stems from the fact that these churches have preserved the thought-forms and religious concerns that matter to the African people. It is these thought-forms that give colour and meaning to the Christian gospel and its message as it is preached (Fashole-Luke 1976 and Mbiti 1972).

An area which has not been given adequate attention, and which would be useful for this essay, is the whole context within which new religious movements emerge. Here again we would find the basic Biblical revelation in juxtaposition, with the "living" interpretations and articulations of the people as they experience the situation. We get glimpses of this in Martin's (1964) accounts of the emergence of messianism in Southern Africa. Socio-economic and political subordination in the Eastern Cape and in Namibia, for instance, made the people to be ready to try a "new religion". The appearance of the comet in 1910 and political dissatisfaction, as well as the gathering clouds for World War I, provided another incubation climate for new religious movements.

Similarly, the 1918 influenza epidemic and the drought and depression in the 1930s were upheavals in the society. Their interpretation and the way to resolve them required an appropriate kind of genius. The Bible and indigenous spiritual leadership did not play a small part in addressing the need.

Conclusion

The relation between Biblical revelation and extra-biblical revelation is a field of study which merits attention, particularly if the development of the new religious movements is going to be considered meaningfully. The Bible with its content of revelation occupies a pivotal position in the Indigenous African Churches. This happens in spite of the fact that much of the membership is illiterate. While in literate societies the Bible would have an independent existence where its creation would be independent of its transmission, it is different in the indigenous churches. The Bible triggers off spiritual transcendentalism. In the process the prophet and religious leader establishes a valuable connection between transmission and existence. At the same time the congregation is involved in the actualization and the creation of the message.

The co-existence of Biblical revelation and extra-biblical revelation in the African indigenous churches extends the concept of revelation which, for a long time, has been limited by a christocentric approach to the availability of the grace of God. This essay has attempted to suggest that in the indigenous churches the issue revolves around helping the full growth of Christ which means the continuing incarnation of the Son of God and the revelation of His true name in deed and message. This does not, however, happen without its own difficulties which stem from the level of literacy of the inquirers.

BIBLIOGRAPHY

Baillie, J 1956. *The idea of revelation in recent thought.* New York: Columbia University Press.

Barrett, D B 1968. *Schism and renewal in Africa: an analysis of six thousand contemporary religious movements.* Nairobi: Oxford University Press.

Fashole-Luke, E W 1976. The quest for African Christian theologies, in Anderson, G H and Stransky, T F, *Third World theologies.* New York: Paulist, and Grand Rapids: Eerdmans.

Küng, H 1967. The world religions in God's plan of salvation in Neuner. J (ed). *Christian revelation and world religions.* London: Compass Books.

Lea, A 1925. *The native separatist church movement in South Africa.* Cape Town: Juta & Co.

Martin, M L 1964. *The Biblical concept of messianism and messianism in Southern Africa.* Morija: Moija Sesuto Book Depot.

Makhubu, P 1988. *Who are the Independent Churches?* Johannesburg: Skotaville.

Martin, M L 1971. *Kimbangu: an African prophet and his church.* Oxford: Basil Blackwell.

Mbiti, J S 1972. Some African concepts of Christology, in Vicedom, C F, *Christ and the younger churches.* London: SPCK.

Neuner, S J 1967. Introduction, in Neuner, J (ed). *Christian revelation and world religions.* London: Compass Books.

Pannenberg, W (ed). 1968. *Revelation as history.* New York: Macmillan.

Rahner, K & Ratzinger, J 1966. *Revelation and tradition.* London: Burns & Oates.

Schwarz, H 1985. *Divine communication.* Philadelphia: Fortress Press.

Smith, E W 1936. *African beliefs and Christian faith.* London: The United Society for Christian Literature.

Shorter, A 1983. *Revelation and its interpretation.* London: Geoffrey Chapman

Turner, H W 1967. *African independent church,* Vol 1: *History of an African independent church: the church of the Lord (Aladura).* Oxford: Clarendon Press.

Turner, H W 1971. A new field in the history of religions. *Religion: Journal of religion and religions* 1 (1), 15-23.

Turner, H W 1974. Tribal religious movements, New *Encyclopedia Britannica* Vol. 18, 697-705.

Turner, H W 1977. The primal religions of the world and their study, in Hayes, V C (ed). *Australian essays in world religions.* Bedford Park: Australian Association for the study of religions.

Turner, H W 1979. *Religious innovation in Africa.* Boston: G K Hale & Co.

Walls, A F 1976. Africa's place in Christian history in Probee, J S (ed). *Religion in a pluralistic society.* Leiden: E J Brill.

PART FOUR

ON UNDERSTANDING, SELFUNDERSTANDING AND MUTUAL SHARING IN AFRO-CHRISTIAN CONTEXT

CHAPTER 13
IN SEARCH OF UNDERSTANDING: ANTHROPOLOGICAL FIELDWORK AMONG ZULU ZIONISTS IN KWA MASHU.

by
Matthias Mohr

"We would like to.....encourage people from various disciplines to study not only a most interesting development in Southern Africa but a most meaningful religious movement for those who became part of it"
(G C Oosthuizen 1986).

The trademark of anthropological research is its diversity. Subjects like history, psychology, musicology, sociology, linguistics and even medicine all make their separate contributions to our understanding of human behaviour and interaction. This means that an anthropologist can seldom be a narrow specialist. Whatever the particular interest of the researcher, he/she also has to be a part-time historian or linguist, for example. Yet, the study of religion is perhaps the most demanding field that an anthropologist can enter, because of its complexity and its importance for every individual adherent.

However, that does not necessarily imply that, in order to study a form of religion and its adherents, one has to be a part-time theologian. The main reason for that is that theologians preferably study their own religion. It is, therefore, natural that theologians, when studying a religion different from their own, apply their own values to it and are concerned with the extent to which the religion being studied can meet the standards of their own system of beliefs (cf

Greschat 1987:129ff). Theologians can study other religions only as polar reflections or outright distortions of their own. The approach is inevitably judgmental and leads to an evaluation of the shortcomings of other religions by the yardstick of one's own. This inability to see what others are doing except as deviation from one's own definition of truth is ethnocentrism. It is by no means confined to theologians, who are at least honest about it. But an anthropologist must avoid it like the plague.

In opposition to theology is the science and history of religion, which is not ethnocentric. Characteristically the science of religion researches foreign, indeed all, forms of religion from an equal distance. A major aspect is to approach the object of study in a comparative way, in order to shed light on its historical and contemporary phenomena. Rather than measuring a certain form of religion against, let us say, Christianity, the science and history of religion seeks to be a science of understanding (cf Mensching 1981:12 ; Geertz 1969:41). Of importance here is the living experience of religion and the ways in which it serves the individual and the group (Geertz 1969:40; Mensching 1981:284ff).

This comes closer to the social anthropological approach if we follow Evans-Pritchard's conclusion that people and their religion have be to studied in action (1965:120). Thus, in order to reveal and understand any form of religion, comparative and relational studies are imperative (p.120). Also, if we agree with Jensen's suggestion that a person is a performing being (1951:405), it becomes clear that, through the anthropological task of studying people and their behavior, we come nearer to an understanding of a religion if we first learn about the people who practise it. It is in emphasising this fundamental understanding that the 'science of the human' (Anthropology) differs from the science of religion.

The understanding of the human dimension certainly presupposes a cooperative application of the concerns of religious science and history, hence making the anthropologist a part-time religious historian and scientist. However, his primary concern is with the human and social dimensions of religious experience. Nevertheless, there is sufficient common ground between Anthropology and the Science of Religion to promote meaningful collaboration between them in the study of religious phenomena.

It is to his credit that Professor G C Oosthuizen has always understood and appreciated the contribution that anthropology and other specialised disciplines have made to the field in which he has distinguished himself, i.e. the scientific study of religion. He has been pre-eminent among those who promoted and engaged in interdisciplinary collaboration and scholarship concerned with the rise and spread of the African Independent Churches (AIC) in South Africa. Most of what has been achieved in this area of study owes much to his motivating efforts and unfailing enthusiasm. In sympathy with his approach, and partly in defence of it, I see it as my task here to set the record straight regarding the part played by anthropology in the study of the AICs. In particular I like to address some recent criticism regarding the validity of the anthropological approach to this topic.

The source of criticism is Paul Makhubu, the leader of an AIC of the Ethiopian rather than the Zionist kind. Since he has published his reservations, these might easily, though mistakenly, be taken to represent the opinion of most, if not all, AIC leaders. For this reason alone, Makhubu's criticisms must be taken seriously. Nonetheless, they rest on a misunderstanding of what anthropologists, and perhaps other scholars, set out to do when studying religions and religious organisations. In his recent book (1988), Makhubu expresses the fear that outside scholars, such as anthropologists, are not motivated by a genuine sympathy ('sincere love') for and interest in the Independent Churches; that their only interest is to use the churches as a quick route to academic preferment. Moreover, in the course of a recent academic symposium, he also expressed doubts whether my ongoing study of Zionists and of the level of their political consciousness, could reflect the real situation. His scepticism was based on his belief that no anthropologist, who happened to be white, could hope to be told the truth by black people. Leaving aside the mere alarming implication that no meaningful dialogue is possible between blacks and whites in this country and that the future is consequently bleak, I wish to assure Rev Makhubu that his pessimism regarding the validity of such exchanges is misplaced, certainly so in relation to the way in which anthropological research is conducted. My response, which is not advanced primarily as a form of personal justification, is to reflect upon my own fieldwork experience. It will do much to eliminate confusion and to clarify the situation, if Makhubu's sincere doubts can be placed squarely in the real context of actual fieldwork research. To this end, I intend to discuss the *aim* and *implementation* of my anthropological fieldwork among Zulu Zionists in the township of Kwa Mashu, near Durban, which I completed after almost two years in March 1992.

What I mean by *aim* here is not the proving of an hypothesis, but rather the personal goals which drive an anthropologist to study Zionists, and the rewards and satisfaction that one gains from it.

When one picks a research topic it is usually not the outcome of a random choice, but the result of previous studies and the influence of a certain personal interest. Before embarking upon the actual research , some kind of hypothetical structure will have emerged in one's mind, as a result of literature consulted.

In the case of the Zionists in South Africa, a large corpus of good quality literature exists, and authors like Sundkler (1961;1976), Kiernan (1974-90), West (1975), Williams (1982) and Comaroff (1985) have provided extensive information about these churches and their adherents. Thus, one can gather a rather good insight into Zionist life, history, tradition, ritual and organisation. However, at this point it would be premature for a researcher to come to any final conclusion, purely on the basis of the literature.

There are two issues which support this assumption. Firstly, literature which is based on extensive fieldwork naturally imparts the author's personal experience, which might be influenced by various circumstances. Therefore, in order to get a somewhat comprehensive grasp of Zionist views and rituals, *own* fieldwork seems to be an appropriate method. Only then one is able to conduct a profound comparison with the existing literature and sources. Moreover, inevitable gaps exist in the literature. Because of the particular interest of authors some aspects will be researched more profoundly than others and some may receive inadequate attention. The advancement of knowledge and understanding often requires one to fill such gaps. In addition, every existing study is contextualised regarding time and place. As these circumstances change, the conclusions reflected in the literature need to be tested and updated by further research. It was a combination of these reasons that urged me to focus on the degree of political awareness among the new generation of Zionists.

Secondly, a final conclusion also depends on our understanding of the Zionists as creative agents and, most of all, on what they can tell us about themselves. Unequivocally, final conclusions should aim to analyse the results of such empirical research within a given context, and not attempt to characterize Zionists as a whole. A non-Zionist will struggle to understand Zionism in all its complexity. Yet, it is precisely this struggle that is embodied in the study of religion, and of the Zionists in particular: many things are unpredictable, and reaching for our aims we have to expect sudden turns and detours, occasionally even dead ends. In other words, the challenge of such a research distinguishes itself through an enduring process of interactive learning rather than through the employment of deductive reasoning from what is already known.

The process of learning is a permanent companion during the process of the *implementation* of the fieldwork. Again, certain expectations rule the first trips into the field, but again, the reality is different.

In the beginning different anthropological fieldwork methods, such as participant observation and various interview techniques (cf. Malinowski 1966:7ff; Pelto 1970:90ff), are taken into consideration. Hitherto, my experience has shown that theoretical methods cannot be applied to solve the problem. Once more the learning process is activated when we adapt to the environment of our fieldwork, and from there gradually mould our own research method - most likely a miscellany of different approved methods.

Adapting to the fieldwork conditions is strongly linked with the researcher's attitude towards the people studied - in our case the Zionists. As mentioned above, Makhubu (1988) feels that a "sincere love" of the AICs would more or less guarantee a balanced study of its adherents. Yet, how can I "love" something I do not even know yet? Would an affectionate approach not hamper the liberty of action? I suggest that instead of "love" we use the word "sympathy" which may be interpreted as an openness, geared towards understanding the truth as others see it.

I must admit that it is anything but easy to maintain this openness and to repress existing prejudices. Especially during the first visits into the field, I could not help feeling wary and insecure. What would it be like to work in a black township? News reports and many white Durban residents painted a rather unfavourable picture of Kwa Mashu. How would the Zionists receive me? Would I be considered as an individual or as a white intruder who was merely 'sniffing' around? Initial research is plagued by such uncertainties.

Whenever I met Zionists I was expected to explain the purpose of my inquiries, since religious values and behaviour are very personal and intimate matters, certainly not to be given away to somebody who appears on the doorstep on a Saturday afternoon. Further discussions about political convictions of the church members evoked an even more sensitive reaction. Fear and distrust were the natural obstacles I witnessed during most of the early interviews. Therefore, allowing for a period of getting to know each other is essential. That period can take two weeks, six months or even longer. Accordingly, my fieldtrips were and still are constantly aiming at constructing new relationships, building them up and cultivating friendships. The latter cannot be done by randomly visiting one person or church after the other. Regular and purely social visits to respondents are of great value.

The social visits in particular are likely to reveal behaviour and events which would not surface during an "official" visit. I was thus able to obtain important information while I was having tea and biscuits or simply watching soccer on television with some of my respondents (more accurately, friends) Such gatherings, I believe, obliterate or at least diminish the status difference between researcher and respondent. Both sides feel more comfortable, and the initial hurdles of conversational constraint may be overcome. There one is able to see Zionists being Zionists *outside* the formal arrangement of their church. Zionist conceptions of certain values start to make sense when the adherents are met during their everyday life. Formalities are mostly put aside and many Zionists made me feel very much at home, and I was accepted into their family circle. Through such close relationships a mutual trust is negotiated and becomes invaluable in extending and securing further contacts.

Finally, I must give credit to Makhubu and other concerned AIC representatives for their scepticism. No doubt, it derives from an historical legacy in which the African Independent Churches were looked down upon by mainline Christian bodies of predominately white composition. But I am not so arrogant as to expect to know the full truth about the Zionists, once I have finished my research. This, I believe, is for them to decide. I am also aware that in some cases respondents deliberately withheld personal information and insisted on their privacy, which is their privilege to do.

However, I am convinced that every little step to learn about a different culture and its different practice of religion will finally lead to mutual understanding and perhaps friendship - both desperately needed in South Africa today. The bridges of the future cannot be constructed on the animosities of the past.

ACKNOWLEDGEMENT

I would like to thank Professor J P Kiernan for the supervision of this article. His valuable comments and advice were of great help. Any shortcomings in this article are the author's responsibility.

BIBLIOGRAPHY

Comaroff, J 1985. *Body of Power. Spirit of Resistance - The Culture and the History of a People.* University of Chicago Press.

Evans-Pritchard, E 1965. *Theories of Primitive Religion.* London: OUP.

Geertz, C 1969. Religion as a Cultural System in: M. Banton (ed.) *Anthropological Approaches to the Study of Religion.* London: Tavistock.

Greschat, H J 1987. *Was ist Religionswissenschaft?* Stuttgart: Kohlhammer.

Jensen A E 1951. *Mythos und Kult bei Naturvolkern.* Wiesbaden: Franz Steiner Verlag.

Kiernan, J P 1974. Where Zionists draw the line: A study of religious exclusiveness in an African township in: *African Studies*, 33 (2).

------ 1977. Poor and Puritan: An attempt to view Zionism as a collective response to urban poverty in: *African Studies*, 36 (1).

------ 1990. *The production and management of therapeutic power in Zionist churches within a Zulu city.* Lewiston: Edwin Mellen.

Makhubu, P 1988. *Who are the Independent Churches?* Johannesburg: Skotaville.

Malinowski, B 1966. *Argonauts of the Western Pacific.* (7th ed.) London: Routledge & Kegan Paul Ltd.

Mensching, G 1981 *Die Weltreligionen* Wiesbaden: Drei Lilien Verlag.

Oosthuizen, G C (ed.) 1986. *Religion Alive.* Cape Town: Hodder and Stoughton.

Pelto, P 1970. *Anthropological Research: The Structure of Inquiry.* New York: Harper & Row.

Sundkler, B 1961. *Bantu Prophets in South Africa.* London: OUP.

------- 1976. *Zulu Zion and some Swazi Zionists.* London: OUP.

West, M 1975. *Bishops and Prophets in a Black City.* Cape Town: David Philip.

Williams, C 1982. *Ritual healing and holistic medicine among the Zulu Zionists.* unpublished PhD thesis, American University, Washington, DC.

CHAPTER 14
"ZION".
ZIONISTS DOING THEOLOGY

by

D C van Zyl

The South African Department of Planning, Provincial Affairs and National Housing has about 7500 files containing particulars of so-called African Independent Churches. In 3252 of these the word "Zion" occurs in their official names (Retief 1991:3). By February 1992 I have recorded almost 100 different AICs in the Cape Peninsula - and there are many more. Of these some 80% have "Zion" in their names. In most cases a clear distinction between Zionist and Apostolic churches cannot be made. Only a few of the local "spiritual" or "holy" churches which do not have "Zion" in their names, claim not to be Zionist. I could thus far locate only two Ethiopian churches in the Cape Peninsula.

The historical roots of the term "Zion", which is so widespread throughout sub-Saharan Africa, seem to be undisputed (Sundkler 1970:48;54; Daneel 1987:54). The question arises as to what the members of these churches themselves understand by this term. A number of "spiritual" church leaders in the Cape Peninsula were interviewed on the question of what "Zion" means to them. The general answer to this question was that it refers to the "umzi kaThixo" (house[hold] of God). Some explained that by calling these churches "Zionist", they confess that these are the churches where God is truly present. The reference to the Old Testament temple and to Mount Sinai plays only an indirect role in their understanding of this concept. The emphasis is on the present: these churches are the "imizi kaThixo".

It is often found that the head office of a specific AIC is called "Zion/Zion City" (Sundkler 1970:292; Daneel 1987:25;103-106 et al). Not a single respondent interviewed knew of a Xhosa-speaking AIC in the Cape Peninsula, Transkei or elsewhere which has a head office called "Zion". "Zion" seems to mean to them the local congregation.

In my inquiries I came across an exceptional and highly informative perspective on what "Zion" represents. This view is held by two bishops whom I happen to know quite well over a period of time. They have close links and a long standing friendship. They have discussed this issue among themselves before. Since they are in agreement on most points, they seem to have influenced one another to quite an extent. They are, however, not aware of anybody else who holds similar views. I could not trace anything of the same nature, not in field studies nor in written sources.

According to the views of these two bishops, "Zion" refers to the two stone tablets written by God and which were kept in the ark of the covenant. I Kings 8:1 forms the cornerstone of their interpretation. This text reads as follows: At the dedication of the temple they "..brought up the Ark of the Covenant of the LORD from the city of David, which is called Zion" (NEB). The Xhosa translation, on which they base their interpretation is ambiguous. According to this translation "Zion" could either refer to "the city of David" or to "the ark". In fact, one is tempted to read it as if it refers to the ark. "yiZiyon" seems to refer to the ark, since "yi-" refers to the noun class of the ark. In fact it is a short form of the proper "uyiZiyon", which would have put the matter beyond doubt. Because of the shortened form, the demonstrative pronoun changed from "lowo" to "leyo". This adds to the confusion. Their interpretation of Zion as referring to the ark is therefore fully justified and logical. In verse 9 of this same chapter it is explicitly stated that the contents of the ark were the two tablets of stone ("amacwecwe") on which the Ten Commandments were written. The special importance of these were that they were written by God Himself. These bishops seem to take it that it was written before Moses "found" it on Mount Sinai, even perhaps by the time of (or even before?) creation. "Zion", therefore, refers to the tablets inscribed with the Ten Commandments contained in the ark of the covenant.

According to I Kings 8:10-11, the glory of the Lord descended on the temple only when the ark was placed in it. It is, therefore, the ark and its contents which make the temple and the mountain to be called "Zion". When asked whether II Samuel 5:7 does not suggest that the mountain/stronghold was called Zion before the ark was brought to it (see 6:16-17), they remarked that it was called so in advance by way of a prophecy. The real aim of conquering the stronghold was to settle the ark there. One of the bishops stressed that, according to the Xhosa translation, the stronghold was not called Zion, it is "the stronghold of Zion", that is the ark. On the question since when the ark was called Zion, since this text in II Samuel is the first in the Hebrew Bible to mention Zion, Isaiah 51:16 was quoted. According to them, this verse which reads "..that I might fix the heavens in place and form the earth and say to Zion 'You are my people'.." indicates that the name "Zion" was known since creation.

Next they quote Psalm 87:2, which says that "God loves the doors of Zion". This shows the supreme importance of Zion. The doors in Xhosa reads "amasango", "entrance", and refers to the way of admittance to Zion. Since Zion does not mean the house ("indlu" =building) but the household ("umzi") of God, it refers to the way one becomes a member of the household of God. This is, according to them, exactly what verse 5 of this psalm speaks of when it says: "to originate from Zion, it will be said: this one and that one was born in it" (literally from Xhosa translation). It means that it is only those who were born of Zion who are true believers, namely the Zionists. Therefore, it is logical to them that Isaiah 51:16 (quoted above) can call the people of God: "Zion".

The next step in their argument takes up Isaiah 40:9, which reads as follows in the Xhosa:

> "You, Zion, preacher of the good message, go up on the high mountain; you, Jerusalem, preacher of the good message, lift your voice with power..." (translated DvZ).

Here Zion refers to Jesus, for He is the one who brings the good message, He is the one who went up the mountain to proclaim the new law of the Kingdom (Matthew 5:1) [In some Zionist services the Beatitudes serve the function which the Ten Commandments do in some major Protestant liturgies]. This, they say, is clear because the Old Testament must be interpreted as speaking of Jesus Christ, quoting Luke 24:44-47. The synonymous saying referring to Jerusalem is likewise interpreted as referring to Jesus, "Ulikhaya lethu lisezulwini" (He is our heavenly home).

This interpretation of "Zion" is concluded in full New Testament perspective. II Corinthians 3:13 is quoted. Now the law is no longer written on stone tablets carried in the ark. God has written his "incwadi" (letter, book), taken to be His (10?) Commandments on the tablets of the hearts of believers by His Spirit. In this way "Zion" is understood as the life of every believer, every baptized member is Zion/ a Zionist. This is what I Corinthians 6:19 means when it says that the body of a believer is a temple of God's Spirit.

The essence of "Zionist theology" is now also interpreted within this framework. To become a believer/Zionist, one has to enter through the "door" (Psalm 87:2, see above). Of course, Jesus is the door (John 14:6), He is the preacher of the good tidings (Isaiah 40:9). Therefore one has to have Jesus, to believe in His Name. But faith without works is dead. The first essential "law" which accompanies faith is baptism (Mark 16:16). Baptism is the door, the door which leads to the river. The door is still closed to those who have not been to the river (Nahum 2:6). Every believer must enter the river, like Jesus himself (Matthew 3:19) and the Ethiopian (Acts 10:47). Baptism must be practised in rivers or the sea, because the Holy Spirit dwells there. Not only did the Holy Spirit come to Jesus when He was in/under the water and did it enter the water, the Spirit of God hovers over all "big waters" from the beginning (Genesis 1:2). The other essential items which put faith into actions, are the garments (Mattew 22:11-14), which should be white (Revelation 6:11 et al). One of these bishops wears only a pure white garment. The other one, who is president of an umbrella organization of mostly Xhosa-speaking AICs, wears a uniform: a dark suit with an official banner and white gloves.

These, baptism in running water, special garments, along with the use of water in healing and purification ceremonies, were mentioned by most Zionists interviewed as the distinctive marks of Zionist Churches. A few respondents who claimed to be Pentecostal and not Zionist, mentioned the fact that they do not use water in healing ceremonies as proof that they are not Zionists. However, one of the bishops mentioned above, is of the opinion that the use of water is not customary to Zionists. He himself uses water only when the Spirit tells him to do so in a specific case.

The case quoted here reveals some important features of the way in which theologically untrained Zionist leaders "do theology". The most important and significant feature in this case is that they managed to construct an almost complete theological system --in terms of the essential elements of Zionist theology at least. Secondly, the role that the Biblical text plays is very important, as is clear from the foregoing discussion. Of course, texts are interpreted within a more or less pre-fixed frame of mind (cf. Psalm 87 and Nahum 2:6), but few theologians within other theological traditions manage to escape this trap! Sometimes texts are spiritualized (cf. "doors" in Nahum 2:6, which is a judgment saying against Nineveh). In other cases strictly literal interpretations are followed (eg. white garments in Revelation). What is more is that this whole interpretation of "Zion", as discussed above, was sparked off by a very close and authentic reading of the Biblical text, in this case I Kings 8:1. Lastly, the thorough christological reinterpretation of the Zion concept and the way in which it is made relevant to touch the life of every believer, is highly significant.

This understanding of "Zion" is indeed exceptional and not likely to win much field. What is of importance is that it shows a real effort to be faithful to the Biblical text and to work out detail in a "systematic" way. Many who hold Zion to mean "umzi kaThixo" or refer to Old Testament institutions, cling to a tradition which they cannot explain and which has very little significance for their daily life.

BIBLIOGRAPHY

Daneel, M L 1987. *Quest for Belonging.* Gweru: Mambo Press.

Retief, E F 1991. The AIC - their Church Names and Listing of Church Names. Paper presented at the Nermic Conference at the University of the Witwatersrand, 4 July 1991.

Sundkler, B G M 1970. *Bantu Prophets in South Africa.* London: Oxford University Press.

CHAPTER 15
EXPLORING FIRE AS FORCE
IN A CONTEMPORARY AFRICAN SETTING

by

A.S. Van Niekerk

When, in ancient Hellas, Heraclitus added Fire as fourth cosmic element to Earth, Air and Water, he was articulating an awareness of the great, also psychic, significance of fire over the ages for mankind, in Africa as elsewhere.

In the industrialized West this mythical, magical aspect of fire has to a large extent been overlaid by a predominantly rational and technological emphasis.

As a young minister in Venda in 1978 - 1983, I soon encountered vestiges of that other, deeper involvement with fire and, I must confess, was at a loss to interpret and deal with it.

At a conference, the youth of our congregation would, for example, all through the night clap, chant, dance, and sing over and over again to the same words.

> Ke koloi ya Elia
> Ge e duma, e a sepela
> (The wagon of Elijah
> When it roars, it goes)

What hypnotic power was at work here, and what role did fire play in it?

One thing was clear, one was dealing with something immensely powerful, an image associated with the godhead itself : "The Bavenda speak of God manifesting himself in a mountain, as a great flame accompanied by the sound of clanking irons" (Mbiti 1971:149).

Now, many years later, an understanding of what fire means to the largely pre-industrial rural African, has become one of the great questions with which our multi-disciplinary research team is grappling. This team, of which I am the project leader, aims at interpreting the perceptions of the voiceless poor to those who make decisions about development projects for them.

Again and again when the Western mind is baffled by Black response to developmental and other issues, we come up against our lack of understanding of the symbolic importance that fire appears to have for the African psyche. We are beginning to realize that it may hold an important key to subliminal forces which we dare not ignore.

Prof G C Oosthuizen was one of the first to grapple with the symbolic meaning of fire in African culture and the AICs. He has through continued research contributed greatly to a better understanding of these matters since he wrote, nearly thirty years ago, that the symbolism of lightning in Shembe's thinking is a return to traditional Zulu thought patterns: "... for the pre-Christian Zulu lightning is 'heaven' in a concentrated form while for the post-Christian Zulu, if lightning is 'thrown' at a person he is filled with the Holy Spirit. Lightning is God's direct power working without an intermediary and so the Holy Spirit is thought of as God's power" (Oosthuizen 1965:142).

1. FIRE AND HOUSEHOLD TECHNOLOGY

In Venda some of us, ministers of the church, became interested in the role of fire in the daily lives of our church members and the community. Women prepared food over open fires, and they spent considerable time and energy to collect firewood and to carry large bundles over long distances on their heads.

In the densely populated areas that were developing everywhere, no firewood was available within walking distance. People bought loads of firewood from vans, at prices that few could afford, and the cutting down of trees extended beyond walking distance into surrounding areas.

The open fires in the household clearly did not make economical or ecological sense anymore, and we thought we would solve the problems by introducing new technology : a "haybox"; a mud-stove that was used elsewhere in the world; a solar oven. These gadgets could save 50-100% on fuel, and we expected that the poorest households would show great interest, and quickly adapt to these in order to save them considerable time and energy. There was, however, hardly any interest in our ideas.

Why? We suspected that it had something to do with the meaning of fire, the emotional relation to fire, the symbolism. While I was visiting church members, I watched the old mothers preparing food. It was not a chore, it was a ritual : the adjustment of the burning wood, the smoke, the care for the food. Here all things come together: the food that was produced on the fields, the firewood from the bush, the water that was carried from the river, the family that is gathering from everywhere to come home, to share food, to be nourished. It has been like that for endless generations, a primordial experience. I suspected that the ancestors, also called the shades, must also be present somewhere: maybe in the flickering shadows of the flames, and in the smoke.

Clearly, people would not accept new technology that did not satisfy these deep emotional experiences. The symbolic and emotional dimensions of fire had to be *included* in the design of new technology, if people were to accept it.

We then started to form a multi-disciplinary research group to see how all the different factors combine in the household: technological, economical, social, religious, emotional - and how better solutions could be designed for the poorest households.

The design of a better stove could also benefit society at large. Black cities are choking under the smoke spewed out of countless chimneys; in rural areas deforestation turns Africa into a desert of soil erosion. Many efforts to provide a cheap, smokeless, effective stove for the polluted cities as well as the rural areas have so far not achieved much, because, we believe, the symbolism of fire has not been accounted for.

In 1989 the Development Bank and the Solar Energy Society of S.A. held a conference on Alternative Energy Supply Options of Developing S.A. At this conference it was said that the consumption of firewood had already surpassed the rate of reproduction of wood, leading to dire ecological consequences. While a properly designed stove could be "a tool to combat the degradation of our forests", those that had been developed had not functioned as they were supposed to. Further research was still required. There were substantial amounts of unused sawdust residues which, according to estimates, amounted to 300 000 m^3 per annum in 1985. It has been estimated that these resources were large enough to substitute the current wood fuel required.

We were told about a stove that had been designed by Mr Fred Hose of the CSIR, and which could work with sawdust. We decided to adapt it and re-design it until the poorest families would accept it as something that satisfied their emotional, financial, social, and other needs: a stove that would really be a tool to improve the quality of life. We placed two stoves with two families and started to study their functioning.

We asked people about the presence of the ancestors, about the flickering shadows and the shades, the smoke - but we could find no evidence, no information that we could use to improve the technology of our stove. Maybe we were searching in the wrong direction?

One thing, however, is becoming clear : the symbolism of fire in any given culture or religion is deeply embedded in the thought patterns and the world-view of that culture. Any new technology that is not structured in such a way that it fits into the patterns and processes of such a culture, will not be readily implemented by the people.

During the seventies, fire has moved into the centre of events in South Africa, often in a destructive role. Again, without a better understanding of the symbolism of fire, a proper response to events is impossible.

2. FIRE AS AGENT OF PURIFICATION THROUGH DESTRUCTION

In several African cultures fire can be used to destroy hostile and evil people: "When snakes have been killed they must be burnt with fire ... because fire destroys them totally. They are rubbish. Snakes and abathakathi (witches) are the same. Both are just rubbish. So that is why abathakathi ought to be treated with fire (i.e. burnt)". When a witch is burnt, his\her shade (spirit) is also destroyed, so that no evil agent remains behind. The total destruction of such evil people have traditionally been regarded as a **good** service to society. Western laws that forbid this tradition are seen as protection of evil-doers who are now free to do what they want.

"It was different in the days of Shaka. Then there was prosperity. That was the time when abathakathi were killed. Shaka killed many. The land was nearly cleaned of them those days ... But today the evil is very great. People die everywhere and there is no food. Where are the cattle of the Zulu people today? Where are the children of the nation today? There is nothing of it at all" (Berglund 1976:305/6). (No children at all? Westerners who see a population explosion in Africa should take note: here is a different thought pattern).

Before 1976 people complained to Berglund that evil people were not eradicated anymore by burning them. In 1976 political protest against apartheid erupted *in which fire played an important role.* Many *buildings* were burnt down in many cities and rural areas. In the 80s the burning of *people* became prominent.

2.1 The burning of property

In 1983 an Afrikaner theologian tried to find an answer to the question why Blacks burnt down *their own* churches and beer halls (König 1983:75). He thought that it was because Blacks were not allowed to participate in the building up of the existing structures. That certainly is part of the answer, but the symbolic meaning of fire should also be considered. Non-involvement does not fully explain burning down.

There may be reasons why children burnt down the beerhalls where their fathers spent their time and money, but burning down has also become an element of the struggle against white domination. In the early sixties the "Afro-American" writer James Baldwin shocked his liberal white friends by pronouncing: "Yes, baby, I mean burn. We will burn your cities down" (Weatherby 1989:199).

In 1981 the black youth of Brighton, England, rioted and started burning buildings and cars. The youth were inspired by the Reggae music of the Rastafarian movement, who preached that the Africans will escape from the white city that will be burned down, and they will go to Africa, like Lot escaped when Sodom and Gomorrah were punished with fire for their wickedness (Frankfurter Allgemeine Zeitung 1981).

And shortly afterwards in South Africa, Johannes Rantete (1984), his bicycle adorned with the colours of the Rastafarians (incidentally, the same as the colours of the ANC), also interpreted "the rebellion", in which many buildings in Sebokeng were burnt down, as a repetition of the conflagration that had destroyed Sodom and Gomorrah. The fire that destroys is seen as an agent of divine wrath and retribution.

2.2 The burning of people

In the eighties the use of the feared "necklace" to burn people who were seen as enemies of the "struggle" received a lot of attention. In Lebowa there were numerous cases of people who were killed and burnt, because they were suspected of sending lightning to kill somebody. In the months after Mr Mandela was released from prison, the burning of people, both political "sell-outs" and traditional "witches", reached its climax. In some towns as many as five people were burnt in one month (Vlotman 1991:35).

In 1986 our research group visited a mother and daughter who had been struck by lightning. The lightning, like a snake on the ground, reportedly chased the mother (Mrs B) round the corner and into the house, where it struck her in her feet. Since then, the fire slowly moved up in her legs, consuming her. The daughter died a few months after the event, because of the heat and smoke that remained in her body. The community then burnt an old woman suspected of sending the lightning. It is a wide-spread thought pattern in Africa: if something goes wrong, somebody is responsible, often in a mysterious way.

Mrs B's younger daughter said that we should not be surprised about the many people who were killed by lightning, and in political murders by necklace, and the many witches that were burnt nowadays. The Bible predicted that the world would be destroyed by the *fire of God's judgement*.

The view of fire as agent of divine punishment is also found in the AICs. In warn-torn Rhodesia in 1978, Bishop David Masuka said that his father had warned that fornicators, thieves, wizards and all the perpetrators of evil would be destroyed by fire. "Even Mutendi preached that all evil-doers would be destroyed. So in these days we see such fire" (Daneel 1988:474).

A member of our research group, Prof Dieter Holm, interpreted the views on fire with reference to the mythology of the ages. In mythology fire is used to destroy the enemy, or rather to transform it.

Mrs M has chopped down and burnt the tree that provided shade for her squatter's dwelling, because, she said, there was a snake that lived in it. This snake, Mmamogashoa, is the mother of all snakes, and she stays in the sea. She flies through the air, and when she comes to land in the tree, she makes a dangerous wind. Therefore, the tree had to go.

In mythology, says Prof Holm, fire, the snake and the tree often re-appear, e.g. in Genesis 3 and in alchemist mythology. (The alchemists used fire in their efforts to turn base metals into gold.) In Genesis 3 man is sent out of paradise, and the angel with the "flaming sword flashing back and forth" stands between him and paradise.

Western man has transformed nature in the fire of the Industrial Revolution. Nature is no longer a threat to him, said Prof Holm. Mrs M doesn't leave "paradise", that is, the mythological world-view. She has not been transformed by the fire of the Industrial Revolution, and neither has Africa.

Some in the group, however, felt they could agree with some families who regarded the modern forces of fire as dangerous. Some factories have high chimneys, with a burning flame that can be seen from afar. The smoke of the Industrial Revolution poisons the air, toxic waste consumes the earth. The ultimate danger of the modern world is a nuclear war, the most terrible fire in human experience. Such a war could perhaps point to the fires of hell: a total destruction that excludes the possibility of a new beginning. The fire created by technology also attains symbolic overtones.

Our group does agree, however, that there are traditional views of fire and modern views of fire. Pre-industrial views of fire are often related to a cyclical concept of time, in which destruction is taken up in a cycle, and becomes purification or transformation.

Such a cyclical thought pattern is, for example, found in the philosophy of the Greek-Hellenistic world. Around 500 B.C. Heraclitus taught that the primal

Being: fire, changes to water, and from water to earth; earth then turns into water again, and water becomes fire. In the end everything is burnt away; it returns to its original form: fire, and from this fire a new *world emerges. The building up or formation of the world is deficiency, the cosmic fire is fulfilment.* God, cosmic fire, universal wisdom, and the human soul go together: the **drier** the soul is, the **wiser** it is.

The Stoics took this philosophy further: Fire is the wise, spiritual, divine force that shapes and moves the world towards its destiny, where everything is consumed in a cosmic fire. The consummation leads to a new world, there is an endless cyclical succession of worlds. The human soul partakes in the divine spiritual fire, and continues to exist after death, but only until it eventually disappears in the primal cosmic fire (Lang 1959:299-300).

This ancient philosophy is still influential in India where it is believed that cremation is a transition to immortality, and that voluntary death through self-immolation leads to the soul's rebirth in heaven. In the 1960s Buddhists burnt themselves in Vietnam : it was not merely political protest against the USA, it was also a religious act, leaving the body behind, abandoning it, so that the soul can enter paradise (Edsman 1987:344).

This positive view of a fire-death is clearly different from the African effort to completely destroy the evil person by burning his body and even his soul. But, given the cyclical thought patterns in Africa, one would not be surprised if evidence is found that the soul of the evil one, having been consumed and purified in the fire, somehow is transformed, and not destroyed.

2.4 Conclusion

Burning down is destruction. It can also be seen as purification (an African tradition), or the punishment of God (Biblical thinking). And, one suspects, an important factor is the cyclical pattern that underlies the old Greek myth of Phoenix: the "fabulous bird of giant size and great beauty, which lived for hundreds of years and was finally cremated. From its ashes another Phoenix arose, or the same one sprang again, resuscitated" (Kirkwood 1959:78). Burning could also be transformation and fulfilment.

3. FIRE AS AGENT IN CREATING LIFE

In Zulu thought patterns and symbolism fire is closely related to sexual intercourse, meat, beer, and to the shades (Berglund 1976: 159, 174, 225 f., 232 f.). "The fire is hot, like a man and woman are hot, especially a woman having the water (i.e. semen) inside" (p. 159). The fermentation of beer shows that it is hot, like boiling water. And it makes a man hot, as does the eating of meat. "That is why the thing of men (sexual intercourse) is the same as the eating of beer and meat. They all involve heat" (Berglund 1976:225/6). And in all of these the shades are intimately involved.

Fire is sometimes made by rubbing one stick in another. One stick is male, the other female (for the Pedi, cf Mönnig, 1978: 117). This image is of ancient origin, and is found world-wide (Edsman 1987:340). The rubbing causes heat, and fire. In the same way, says a Zulu spokesman, sexual intercourse causes heat. They are the same (Berglund 1976:232).

"So the heat in the fire-sticks is just the very same as the heat in the male and female ... When a man sits by the fire warming himself he must think of women. He cannot leave women out. He must think of them ..." He knows that the fire "comes from the work of one stick in the other. So when he knows this, how can he escape thinking of himself doing this thing, when he has done it before with his wife? He knows it and the desire is there in him. That is how we Zulu think of these things" (Berglund 1976:226/7). Even when fire is made with matches, the warmth of the fire reminds a man of sexual intercourse.

In this thought pattern then, life and the procreation of life is closely related to the shades and to fire: heat, sexual intercourse, meat, beer - in all of these the shades are working to create and strengthen life.

Fire or heat may also be related to ritual impurity that must be avoided at certain rituals. Fire that comes from other homesteads, as well as women who menstruate, who are pregnant or who suckle, or men who recently had sexual intercourse may not be involved in such occasions, because there everybody must be "cool" (Berglund 1976:159, 227).

Two terrible monsters, both related to sexuality, threaten to destroy human life in Africa. The first is the population explosion, increasing the number of people for whom there is not enough food, water, land, and work available. The result is that famine and death increase day by day. The second danger is AIDS that is already killing thousands. No program to contain these killers will succeed if it does not reckon with the symbolism of sexuality and fire as life-creating forces, and perhaps utilise the element of impurity in this symbolism. Maybe the AICs can teach us how to communicate in symbolic language.

4. FIRE IN THE AICS

In the AICs fire does not play the central role that water does. As far as I know there is not a "theology of fire" in the literature (cf. for example the "theology of water" in Turner, 1967:147-150).

Fire is often the opposite of water. Fire symbolises God's judgement and the fellow-person's anger, and any dream in which a person escapes from being burnt (such as fire-walking without injury, or falling into fire without burning) can be seen as "victory over evil, pardon from judgement and, by implication, identification with the Church's ministry". Water symbolises and provides escape from fire, it provides deliverance from drought, the successful evasion of the heated, destructive anger of an adversary, and the "cooling off" of destructive sin (Daneel 1988:53).

Pretorius (1977:130-131) has shown that water is a symbol of sexuality and procreation in many AICs. Does that mean that the AICs have transformed traditional African thought patterns? Could their use of these symbols contribute to the developing of a life-enhancing lifestyle in Africa?

5. GOD AND FIRE

The ambivalent symbolism of fire as a life-giving and a destructive power is almost universal. The exception, seemingly, is the ancient Persian and Iranian symbolism of fire as purely goodness and wisdom; fire may not even be contaminated by burning impure substances such as corpses. In the end, fire will purify the world. Evil is associated with the earth and with darkness; heaven and the new earth with fire and light; (see Lang 1959:932-933).

Lists of different types of fire are also common; in rabbinical thinking in the time of the Bible six ways in which fire appears were distinguished (Lang 1959:937); Vedic India had five fires of different origins (Edsman 1987:342). This can be compared to the different types of fire (modern forces and traditional forces of fire), distinguished by the family of Mrs B.

In the Bible, fire never becomes a cosmic force. In the Old Testament fire is used for everyday purposes and for offerings in the temple. It is also used as a figure of speech, e.g. for anger. It does not, however, function as a force in itself. Israel's neighbours' view of fire as a cosmic element is not taken over. Fire is seen in a completely theocentric way as manifestation of the mysterious, terrifying and enrapturing glory of God, and as symbol and means of his judgement. The New Testament, likewise, does not follow the example of

Hellenistic Judaism (Philo) by describing fire as a cosmic element, or as a culture-creating force. The theological tradition of the Old Testament is continued: the theophany is often associated with fire, or fire is the means of God's judgement : God uses fire as punishment; it is a sign of the coming final cosmic fire and the eternal fire of hell (Lang 1959:936-939, 947).

The ambivalent role of fire is, therefore, also found in the Bible, but in a theocentric way : God uses fire for his purposes; fire is never a force that can function on its own.

6. CONCLUSION

The thought patterns and symbolism pertaining to fire are deeply rooted in the world-view, basic values, and religion of a community. All of these have a profound impact on the lives of people. In recent years, for example, many people in South Africa were suspected of witchcraft or political treason, and burnt to death as a result. The damage done by the burning down of property since 1976 can hardly be estimated. Such events often increased, with certain victories for the liberation struggle against apartheid, such as the release of Mr Mandela from prison.

The symbolism of fire is also related to sexuality and two dangers to life in Africa : population explosion and AIDS.

The fires created by Western technology attain immense proportions with both life-giving and destructive potential.

A better understanding of the deep philosophical meaning of fire in the African and Western world-views may help us to better understand the reason why people do the things they do. The role of fire in the AICs merit further study.

The final solution, it seems, lies in the realisation that the God of the Bible is ruling the world, and that he uses fire for his purposes. Fire is not a cosmic force, neither good nor evil; it is an instrument used by God and by people in a world where God rules supreme. One cannot use fire to destroy evil in the world: Jesus Christ has already destroyed the power of evil. God is the judge; he alone can use fire to punish. If we as human beings arbitrarily try to take this work out of Gods hands, by judging and burning people, we become evil ourselves.

BIBLIOGRAPHY

Berglund, Axel-Ivar 1976. *Zulu thought-patterns and symbolism*. Cape Town: David Philip.

Daneel, M L 1988. *Old and new in Southern Shona Independent Churches, Vol III*. Gweru: Mambo Press.

Daneel, M L 1989. *Fambidzano Ecumenical Movement of Zimbabwean Independent Churches*. Gweru: Mambo Press.

Edsman, Carl-Martin 1987. *Fire* in Mircea Eliade (editor in chief): *The Encyclopedia of Religion Vol. 5.* New York: MacMillan Publishing Co., pp. 340-346.

Frankfurter Allgemeine Zeitung, 25 Juli 1981, Nummer 169.

Gaskell, G A 1960. *Dictionary of all Scriptures and myths*. New York: Avenel Books.

Kirkwood, G M 1959. *A short guide to classical mythology*. New York: Holt, Rinehart and Winston.

König, Adrio 1975. *Hier is Ek!*. Pretoria: N.G. Kerkboekhandel.

Lang, 1959. *pur*. (fire) in: Kittel, Gerhard: *Theologisches Wörterbuch zum Neuen Testament*, Stuttgart: pp. 927-953, W. Kohlhammer Verlag,

Mbiti, John S 1971. *Concepts of God in Africa*. London: SPCK.

Mönnig, H O 1978. *The Pedi*. Pretoria: J.L. van Schaik.

Oosthuizen, G C 1965. Isaiah Shembe and the Zulu world view. in Becken, H.J. (ed.): *Our approach to the Independent Church Movement in South Africa*. Mapumulo: Lutheran Theological College.

Pretorius, H L 1977. *Swartman, seksualiteit en sending*. Pretoria: NG Kerkboekhandel.

Rantete, Johannes 1984. *The third day of September. An eye-witness account of the Sebokeng Rebellion of 1984*. Storyteller Series no. 1, Johannesburg: Ravan Press.

Turner, H W 1967. *African Independent Churches II*. London: Oxford University Press.

Vlotman, Bibi 1991. Blaas toordokters geweld aan?. *Rapport* (newspaper) 20/10/1991, p. 33.

Weatherby, W J 1989. *James Baldwin. Artist on fire*. New York: A Laurel Book, Dell Publishing.

CHAPTER 16
MUTUAL SHARING AS LIBERATION OF "MAINLINE" AND INDEPENDENT CHURCHES

by

B. A Mazibuko

Education is often a liberating venture. It was Paolo Freire, the Brazilian educator, who made us aware of the fact that in what is normally called 'education', two things can really happen: either 'domestication' or 'liberation'.

Education as a liberation process involves influences and actions which can set the human mind free to respond to liberating forces. It is, so to speak, a decolonization of the mind of those people whose dignity has been robbed by 'domestication', the authoritarian type of education.

All colonized people of the world have two selves: the indigenous self over which has been superimposed another self by the Western culture. This culture comes on aggressively, riding on technology, economic development, and the Christian belief. By the time it is with us, we realise what cleavage it has brought about, between an 'educated' elite and the masses, between the indigenous self and a new individual self that moves away from the collective sensibility and glories in Western individualism. Hence derives the disinherited personality. Decolonizing the mind, therefore, will only lead to liberation when it is a rediscovery of self.

Decolonization of the mind, however, must also be a way toward the synthesis between the rediscovered 'I am' and the best that the environment can yield, assuming that the duration we want will develop our aptitude to select from the environment as it stands and not only from the past. The rediscovery of self must therefore be through the discovery of one's community, the recognition of the 'we are'.

The significance of this latter dimension can best be appreciated when we look at ourselves closely and realise how the white person's image dominates the environment in which we live. Through advertising, mass media such as television, radio and the newspapers, sources of production and school text-books, we are constantly being bombarded by images of white people from childhood to old age until we adopt these images as models for ourselves. We are bullied by the white man's terminology. He invents the word 'homeland', and we immediately write off the whole rural constituency as backward and irrelevant to urban-based national politics. In this way we ourselves widen the agonizing rift which the white ruling class has created between the urban and the rural population. When the white "liberal" press and the "progressive" parties of South Africa used to talk of power-sharing, they made it quite clear that they only talk of "urban blacks", not the rural peasants, because subconsciously, they have written off the rural reserves, as "homelands", where the Africans must feel at home under a dispensation of their own.

Decolonization of the mind is a therapy for this condition. As long as we are not policy-makers in our own country, we must devise strategies for non-formal integrated and interdisciplinary education. This means, our education must include all the elements essential to our growth as a people and to our intercultural development. Strategies such as these, yet, do not preclude the possible role of an educator as a decolonizing agent even within the present system. The phase of decolonization requires all the drive, the vigour, the ingenuity and the commitment we can muster, because in this process lies an untapped reservoir of moral strength and purpose. For the educator to be effective in this venture, he himself has to undergo a process of re-education.

Committed educators in South Africa, have therefore to arrange a sustained programme of refresher courses for their members along cross-cultural and cross-disciplinary lines. It is quite evident that many teachers are at present incapable of inspiring their pupils because they lack the "academic" resources which always have to be replenished and updated through self-education and collective learning experiences.

Education, Liberation and Traditional African "Humanism"

There is another dimension of education as a liberation process. It concerns who and what we really are as African people who stand alienated from our indigenous self and yet are not sharing the life of the Western personality holding us in spiritual and intellectual bondage.

We need to know ourselves as a people whose traditions are rooted in the traditional "humanism" - u b u n t u - and in the organic view of life, of the world and of the cosmos. We love life. We are not neutrophilic. We like to touch one another physically. We gravitate towards one another because of the strong sense of community. We have a strong collective responsibility. We have a sense of theatre in which self-expression and self-actualization find fulfilment.

We love to protect the elderly. Our traditional understanding of the Supreme Being and of creation and human growth is all-pervasive. It finds expression in our reverence for life, for universal order as manifested in external nature - mountains, rivers, plants and animal life, soil as the life-giving element, the stars and planets, people - all being interconnected and representing a vital force.

Ours is not a mere church-going mentality or Sunday religion, because, after all, people are part of the universe. The Supreme Force controls this universe; it is within it; it is the universe itself. People are the expression, the extension of the Supreme. The rituals we have are not denominational but communal: birth, puberty, initiation school, marriage, death, ploughing and harvesting seasons, all have their rituals as a celebration of life. The ancestral presence is always involved in all these rituals because the ancestors are alive. They are a historical reality. They are not creatures of fantasy, that is, they once lived in the physical form, in space and in time, and they are now the living dead.

The role of the African church

The genesis and the history of the church among the African people are quite ambivalent and show difficult traits. They have both, clear pluses and clear minuses. The Western white missions did not sufficiently try to understand the black personality, black culture and black religion. Thus, in a true sense, the Christian church contributed to the deculturalization, despiritualization and dehumanization of the African personality. Had the church more carefully studied the African world-view, this could have provided a harmonious linkage between the Judeo-Christian and the African worlds. Instead, the Christian church became used as a tool of exploitation. Also the lack of balance between the teaching of the gospel and the lifestyle of the missionaries, traders and colonists caused problems. This was aggravated by the Western philosophy and politics of fragmentation which separated the communities and divided the people not only along denominational but also ethnic lines.

However, there is a different dimension to the Christian church. For the church also played a role in enriching the African values of dynamic love, sharing, and mutual growth through acceptance and participation. It became utilized as an institution of hope and fulfilment. Most blacks would find it easy to agree with C Eric Lincoln in his assessment of the black church among the Africans who were suffering in the diaspora:

> 'The black man's Church was his school, his forum, his political arena, his social club, his art gallery, his conservatory of music. It was the lyceum and gymnasium as well as sanctum sanctorum. His religion was his fellowship with man, his audience with God. It was the peculiar sustaining force which gave him the strength to endure when endurance gave no promise, and the courage to be creative in the face of his own dehumanization.' (The Black Experience in Religion 1974:67).

For blacks in South Africa, the church has been the only meaningful support community in a society out of their own making. Outside the church, no affirmation was provided for the continuity of black community culture. The original African communities, families and tribes, relied upon their customs and religion for the sustenance of their community. In the developing township-culture, however, where people were bombarded with the effect of urbanization, technology and the mass media, the black church became the exclusive place where the community was sustained and a lifestyle cultivated which allowed people to participate in black culture. Worship and the decision-making processes within the congregations played a most significant role in this. The black church developed into a model for the empowerment of the people. It was and still is the black church which keeps before the nation and the world the justification for human rights.

When referring to black churches the African Independent/Indigenous Churches in particular come to mind. These churches grow in number and impact. Already more than one third of the black population of Southern Africa belong to these churches.

As a general rule, the church for blacks is a community centre, an educational centre, a leadership-development workshop, and a fellowship centre.

The church often functions as a place of healing. Above all, the church is a school for personal growth. People whose personal identity is never affirmed in any way that induces or maintains self-esteem, find in the church a place where they are blessed with the greatest possible affirmation of themselves, the affirmation that they are individually the sons and daughters of the Great God. This helps them to know that each of them is an image of God and therefore just as important as anybody else. They learn to know that God revealed himself in his Son Jesus who is both their Lord and their Brother. This therefore gives a great boost to the morale of the down-trodden and the poor.

The church provides help for women whose role is without doubt the hardest and dullest among those who live in rural black communities. The church affirms them. The church teaches them that Mary was a poor country girl but that, through God's mercy, she became the mother of the Lord of all people and all nations. They realise and appreciate that it is only in the black churches that some daring and radical innovations in ecclesiastical polity permit women to hold important offices and discharge function and duties vital to the life of the church. The pioneer in the black church movement was a woman. The very first leader was Donna Beatrice. In black church movements there are female bishops and founders of a high spiritual calibre.

The church makes a decided impact, not only in rural black wildernesses and deprived areas, but also in our largest urban centres. Since the townships are extremely dangerous places in which to live, the most dependable places which will accept even a stranger, are black churches. Because of the security and other benefits which they gain from church membership, most members strive to become models of the behaviour and lifestyle which is normative in the church community in which they are accepted. These people will never find love and acceptance in the hard world of their daily labour wherever they work, but they will find love, acceptance, education of a kind, and practical help and sympathy in their church. Black people in South Africa know that without these living testimonies to the spirit of the church, there would be no future for them. But the question is: Must they face the future on their own? Is there no one to accompany them on their pilgrimage?

Black and white churches on pilgrimage

The Project in Partnership between Black and White, with Roswith Gerloff as its founder and first director, is a story about a people on their common pilgrimage. It cannot be owned by any single organization or denomination. It is a story which tells about people's lives, their beliefs, their feelings, their competencies, their worship, their actions, their joys and their sufferings. Hence, it is the mutual encounter, or the mutual growth, of those who are willing to take a plunge. It is the journey of the white established churches with about a thousand black independent congregations linked to more than two hundred Christian bodies throughout Britain which originated from a great variety of indigenous traditions - African, Caribbean, Black North American and Black British. All these have for the longest time been ignored by white church and academic institutions.

This pilgrimage, which tells us something about the black movements, says equally something about the white traditions which slowly begin to discover the tremendous contribution to renewal their black brothers and sisters can render. They are either afraid of it, or gratefully aware of such a crucial development. Involved here is both the dimension of joy and of pain. This is true because in this mutual encounter and journey together emerge both the cultural arrogance of some and the humility of others. Both sides become shaken by this experience. The pilgrimage is, therefore, a fascinating but also a very risky venture of people who are called to travel together to a yet unknown destination.

This Project in Partnership, now called the Centre for Black and White Christian Partnership, has revealed that in order to bring ethnic minorities and majorities in Britain together, there must be the evolution of a continuing interaction between the dominant and the subordinate culture, or face to face encounter. The culture of the majority as well as those of the minorities are part of the present dynamics in Britain. The observable data of the last thirty or so years show that there has been a very serious lack of attention to these cultural dynamics and their implications, and also to the resultant ongoing relationships between the different ethnic groups, which in fact should form the integral core of British society. There has been also a continuous failure to create those meeting points where the necessary interaction could take place.

The interaction between dominant and subordinate ethnic groups is, of course, a power struggle which we must not overlook, and which involves dominance and survival strategies, both manifested in and maintained by the economic and social order.

From the very beginning, some of the leaders of the black churches in Britain, both of Caribbean and of African origin, saw - perhaps quite intuitively - the urgent need for such an interaction, at least among the Christian churches. Among others are Martin Simmonds, bishop of the First United Church of Jesus Christ Apostolic; Olu Abiola, bishop of the Aladura International Church, and Samuel Owusu-Akuffo, bishop of the Divine Prayer Society. They took as their primary motive the creation of opportunities for education and training among the leadership of the church on both sides. They wished not for the formation of a normal training scheme provided by British universities and colleges, but for a new way which was innovative and leading to mutual growth. They hoped it was to take into account the real life - practical circumstances of the people, aspirations of the churches, and vocational needs of the prospective "students". Hence the training could not be rigidly pre-planned by a university or a college but rather prepared in consultation with the future participants, who would not only come to learn but also to contribute substantially to the life and direction of such a new learning experience.

Mutual sharing was the major aim of this effort. Roswith Gerloff in her first preparatory paper for the venture in 1976, outlined the aim as follows: "To enquire into the methodology of theological education, especially with regard to the communication between the literal and oral cultures as mainly identical with the cultures of the rich and the poor, the oppressor and the oppressed ... To do this in the framework of a programme exploring the implications of the whole charismatic movement, comparing and contrasting black and white examples and examining interracial fellowships ... To sponsor or establish training schemes parallel to the ones initiated by the Christian Institute of South Africa where leaders of independent churches received literary, social and theological education."

Father Abiola clearly affirmed these aspirations when, in *Partnership between Black and White*, he asserted that education and re-education at academic as well as spiritual levels was badly needed by both the white and the black sector of the church in Britain. The black part of the church, he wrote, needed the expertise of the white part, particularly in respect of scholarly biblical teaching, and the white part of the church needed the spiritual awakening of the black church. This could be done in especially and mutually prepared courses, but even more helpfully through the creation of a residential centre where dialogical education could take place, dialogical because both sides would have to benefit from such a venture, and where people could live together in fellowship. He emphasized strongly that the white leadership would need training which helps them as well as the members of their congregations to revive some of the virtues of the early church, such as charity and tolerance. He distinctly felt that white

Christians would have to get rid of a vocabulary of 'them' and 'us', and replace it with an understanding of the entire Church as 'us'.

Abiola obviously believed that the black church as a whole could help the nation to revive the depth of the "Old Religion", or Bible religion, where a sincere spirituality was the norm. For him, the inauguration of such meeting points was apt to lead to the creation of a new birth in the life of the church in Britain.

Pastors and church workers of the black population came to participate in the project in partnership. They came first and foremost from the school of life. They were grounded in the working class milieu of their communities. They were committed to learning not just for themselves as individuals but on behalf of these communities which they represent. Most of them did not have the usual university entrance qualifications. Some of them had obtained educational qualifications in their native countries - the Africans more so than the West Indians - but their qualifications in most cases were not recognized by the British academic system. A few had undergone theological training in fundamentalist Bible schools, Evangelical or Pentecostal, which were the only ones to offer them training, but these institutions were not acknowledged by the British theological institutions. These leaders of people in many ways felt, therefore, inadequate in the British environment, and thus offered themselves as partners in the new enterprise. They wished to receive education and to learn new skills, but they could also contribute from their own expertise.

Roswith Gerloff observed that the majority of members of the project always claimed that the "real university" of their lives had been the local church from which they stem and which is their spiritual as well as intellectual home. These Pentecostal, Holiness, Sabbatarian or other churches from which they came provide an exciting training ground for a theology lived out, felt out, danced out in the very lives of people. Furthermore, education takes place where the

Christian faith has to prove itself, not only in church, but in the factories, storerooms, buses, kitchens, offices and hospitals. From there they bring a treasure of biblical, social and missionary experience which no traditional European university can ever provide.

Gerloff concluded that people whose eyes are primarily fixed on letters, the people of the 'ocular' cultures such as the Europeans, can learn much from the people of the 'oral' cultures where the way of communication is primarily the song, the dance, the shout, the story, the dream and the joke. In her view God is not ashamed of oral patterns of communication; on the contrary, God asks intelligent questions through the black Presence among us.

Black church experience and new imperatives

The church has served as catharsis for black people's suffering in an unsupportive society. Most of the black church's energy, therefore, was often spent in adapting and re-interpreting the white man's religion to the peculiar needs of those who were regarded as second-class citizens. By recalling its biblical roots, this effort became geared toward ownership in a history of exploitation and deprivation. In this way, blacks appropriated the gospel of Christ to their particular situation.

Those who work in the community today, must, therefore, never forget the viability of the church in the history of black people. In countries such as the United States and the Republic of South Africa - not so much yet in the United Kingdom - responsible educators have worked hard to introduce Black Studies into the traditional curricula of schools. This phenomenon has started and developed largely within the institutions of higher learning but it must continue to penetrate the churches and the community at large, to raise a sense of critical awareness. In Black Studies, the church is used as vehicle in its particularity, and the black religious experience in its universality.

The Black Christian experience is a racial dimension of the church universal, but it has been made a racial minority dimension in the American and British contexts, and a racial exclusion in South Africa. The black church has been generally ignored as a serious historical and contemporary religious and social reality in the life of the white churches, and in the education programmes of theological schools related to them.

The thrust in the years to come is that all of us engaged in the task of training people for these multicultural societies, be made aware that our impact will only be felt if we combine our forces and struggle as one mighty army. This means that Black Studies and Third World Studies, and this includes the study of the African Independent Churches, must be introduced into theological schools. It also involves that both religious and secular must unite around the common issues affecting black people and all ethnic minorities in their daily survival experiences. White congregations in the midst of the black community would be well advised to begin to understand the history and the impact of the black church. As there exists a certain degree of freedom which allows for self-expression and human dignity, it is an essential part in the process of liberation, also of whites.

Blacks themselves are rediscovering the concept of wholeness in this approach, and maybe other minorities could learn from them. They know that for too long their minds have been programmed as though there were two readily separated worlds on which the psyche of people is focused: the secular and the spiritual, religion and social politics. This has resulted in an imaginary and unhealthy division, and in the process valuable energies have been dissipated.

There is, therefore, a new imperative or a new strategy. One aspect of this is that all scholars of whatever persuasion or discipline, religious or political, should seek the common denominator which defines the essential suffering of all black people. To this they must bring every resource at their command. We cannot allow linguistic divisiveness or organizational disagreements to occupy an extended amount of time that ought to be given to the actual task of black liberation. We strongly feel that it is precisely to the extent to which we quickly translate our intellectual language and concepts into redemptive actions, that we shall succeed to move from the struggle to victory. As 'The Illinois Black Scholar' once wrote:

> "Black studies must inform all Black students in every school and discipline that Black people are in one condemnation. Hence Black studies require us to struggle in order to find the basis for one conversation. Indeed we call the so-called religious radicals and the so-called secular radicals out of their esoteric ghettoes into the world of suffering of the Black masses. Jesus of Nazareth said, "Those who are not against us are for us". We therefore challenge all Black Studies advocates to engage in intellectual and social intercourse. We must be able to communicate with each other and strategize together, for we are one people who suffer a common oppression."

The second aspect is that the African Independent Churches also need to be better educated and better equipped in their leadership. During the past decade this need has often been expressed by leaders of African Independent Churches themselves. But action has also been taken. Throughout Africa established churches and interdenominational agencies have been operating in partnership with the African Independent Churches to address these needs.

With this we have already touched on the third aspect of the new imperative: all must take hands and join as friends, black and white, individuals and institutions, mainline and indigenous churches. A concerted effort should be made to no longer see the mainline churches as colonializing churches which are foreign to Africa and its needs and hopes. At the same time the African Indigenous Churches should no longer be denigrated as being syncretistic and even unbiblical.

Enterprising projects should be launched so that mainline churches and indigenous churches can learn from one another, strengthen one another and seek the benefit of one another. Above all, all of us should strive to display a new attitude, the attitude of co-operation in an ongoing project: a spontaneous project in partnership. Such mutual education cannot but result in mutual liberation.

PART FIVE

ON ORALITY
IN
AFRO-CHRISTIAN CONTEXT

CHAPTER 17
SHEMBE PREACHING:
A STUDY IN ORAL HERMENEUTICS

by

J.A. Loubser

Introduction

This article is a presentation of the results of a reading of 23 sermons, preached from 16 August 1980 to 6 August 1983 by Londa Shembe, grandson of the first Shembe, Isaiah (1867-1935), who began his own ministry in 1911.

The sermons, originally in Zulu, were recorded on audio-casette, transcribed and translated into English. Although most of the sermons begin with a literary exercise, they function completely within the oral context within which they were produced and delivered.[1] Whereas the oral tradition underlying the Gospels developed over about thirty years, the oral tradition about the deeds of Isaiah Shembe, spans no less than five decades. It therefore provides an insight into what happens to an oral tradition over such a period of time.

Despite some differences, the Shembe followers exhibit most of the typical features of the Independent Black or Zionist Churches (cf Kiernan 1990). In a general sense one may say that the sermons reinforce the framework of the shame culture (as opposed to a guilt culture) within which they operate. They

---

[1] Although the Shembes themselves were literate and some of their associates could read the Bible, the vast majority of followers are illiterates from the rural areas (Vilakazi 1986:117).

also reflect the reaction of a specific group of Zulu believers (the "Nazarites") to the pressures experienced by the socio-economic conditions brought on by a century of contact with modern society. The sermons were delivered in a largely unpoliticised environment (Vilakazi 1986:116).

Understanding the oral hermeneutics behind the sermons, provides us with a key to significant aspects of the Christian culture among South Africa's traditional society.

SYMBOLIC UNIVERSE

Before examining the orality of Shembe preaching, it is necessary to take note of the context, i.e., the Shembe world as reflected in the text itself.

In order to help its believers find meaning in a disorganised universe, religions project a semantic or symbolic universe. It is to this aspect of the context that we shall pay attention. Here follows a brief summary of the main components in the symbolic universe reflected in the sermons:

1. Shembe faith rests on traditional Christian beliefs in God, Jesus and the Bible. These form the loose framework within which the other elements are arranged.
2. The first preacher, Isaiah Shembe, is seen as a messianic intercessor, or mediator, between the spiritual and mundane worlds. The love of Shembe promotes a sense of well-being in the believer while the first Shembe forms a focal point for human motives. His fears are the fears of the group; his victory over opposition is their victory.
3. The sanctuary "Ekuphakameni" ("at the high place") also serves as an important focal point. The sanctuary unites the believers and mediates an encounter with the spirit world.

4. A belief in the spirits as exponents of the spiritual domain reinforces the awe of the transcendental and replaces mundane motives and fears.
5. The nobility of the Zulu moral norms and tradition - as manifested in polygamy, chastity, dances and the wearing of skin skirts - serves to reinforce a sense of dignity and security.
6. Belief in the exalted destiny of the followers of Shembe reinforces group identity and is demonstrated by the wearing of white garments.
7. Belief in prayer as problem-solving device.

ORALITY AND LITERACY

The basic insight of oral theory is that spoken words have no abstract or permanent, visual, connotation. They disappear as soon as they have been pronounced - *verba volent, scripta manent.*[2] Words are therefore perceived to be events rather than things.[3] For the sake of clarity sharp contrasts are drawn, whereas a variety of mnemonic patterns may be used to preserve valuable information. Normally oral communication also employs heavy rythmic, balanced patterns, repetitions and contrasts, extended forms of alliteration and assonance, formulaic expressions, series of standard thematic settings, proverbs and parallelisms.

---

[2] Vansina (1965:xi) symbolically opens the preface to his book on oral tradition with these words.

[3] Cf also Dewey 1989:35 who refers to Havelock 1963.

These phenomena occur in literate cultures as well, but in oral cultures they appear incessantly.[4]

Because a speaker cannot backtrack to edit his text, incongruencies which arise cannot be ironed out by means of an editing process. They are rather circumscribed by subsequent renditions. Every new formulation already calls for another one. It is therefore useless to look for an "original" formula - as deconstructive theory rightly maintains. In the process unnecessary information is continuously being deleted from memory.

The idea of originality and repetition in oral cultures is therefore different from that in modern literate cultures. Originality does not manifest in the invention of new stories or myths, but in developments due to the live interaction between speaker and audience during narration (Ong 1982a:41f).

ORAL FEATURES OF SHEMBE PREACHING

We may now pay attention to specific features found in the sermons of Londa Shembe. The sermons have been numbered according to their chronological order. The specific page in the transcription is indicated by a semi-colon (e.g. "4:5" points to the fourth sermon, the fifth page).

4.1 Non-linearity

There is very little trace of a linear build-up in the Shembe sermons. Sermon 2 may serve as a typical example. Only the broad theme of Shembe discipleship binds the loose elements together. The following elements appear:

(a) Scripture readings from Genesis 13:1ff (the division of Canaan between Abraham and Lot) and Genesis 19:23 (Lot is seduced by his two daughters); thereafter 1 Corinthians 13:8 (love never fails).

---

[4] Ong 1982:34. Cf also Dewey 1989:35 - chiasms, ring compositions (inclusio) and verbal echoing all facilitate memory.

b) Recapturing of how Abraham had left Mesopotamia. Only Lot followed him although he invited all.

(c) A comparison of Shembe's calling with that of Abraham. Shembe left his wives, not knowing where he was going. Some others followed him like Lot had followed Abraham. Among them was a young girl Shabane.

(d) Shabane's later encounter with the preacher.

(e) Excursion on "happy is a person whose heart experiences sadness."

(f) An old woman's story of how she followed Shembe in spite of opposition.

(g) How Londa heard of the old woman's peaceful death.

(h) Encouragement to the audience: "some of you have long been here in this house being engaged in this journey of sadness and tears."

(i) The episode of the Indian who came to follow Shembe.

(j) The story of the girl Nginisikele Mlaba who had a bad influence on the other youth.

(k) The extended story of Joseshina, called Sulile, who deserted Shembe but was called back by a voice.

The other sermons also have this Midrash-like appearance, with the links between the different sections depending on key words, impressions and associations. Like the second sermon, most of the others are open-ended with no specific appeal or conclusion at the end.

That this is typical of oral communication, becomes clear from the work of Walter Ong. According to him writing restructures human consciousness (1982a:78ff). Oral cultures can produce highly complex thought patterns, but they cannot analyze "elaborate concatenations of causes into the analytic kind of linear sequences" (1982:57). A sober linear or analytical thought pattern is an artificial product of a literate culture. This is one reason why the literary concept of a plot (as found in novels and detective stories) only developed after many centuries of literacy.[5] Oral compositions rather make use of modifying repetitions, which Havelock described as follows:

> "Oral compositions will avoid sheer surprise and novel invention, ... the basic method for assisting the memory to retain a series of distinct meanings is to frame the first of them in a way which will suggest or forecast a later meaning which will recall the first without being identical with it."[6]

This is exactly what happens in Shembe preaching.

2. Oral composition

From the outline of the second sermon in the above section, we can gather that Bible passages are read, but that there is no exegesis of the text. The literary passages merely serve as a starting point for the oral arguments, parables, legends and exhortations that follow. This throws light on the oral composition of the material, which can be described as "a stitching together" of the diverse components.

Milman Parry showed that this is typical of oral communication. In his study of Yugoslavian bards he found that their songs, though keeping the same metrum, were never sung exactly the same way twice. The same themes and formulae occurred, but they were "stitched together or 'rhapsodized' differently in each rendition even by the same poet, depending on audience reaction, the

---

[5] This is why no one had written a tidy detective story before Edgar Allen Poe's "The murders in the Rue Morgue" in 1841 (Ong 1982:144-5).

[6] Havelock 1984:183 as quoted by Dewey 1989:39. Cf also Boomershine's description of orality in the first century (1987:145f).

mood of the poet or of the occasion, and other social and psychological factors" (Ong 1982a:59-60). It is this same "rhapsodizing" that form the basis of the Shembe sermons.

3. Time conception — a-historicity

In studying the sermons it soon becomes apparent that biblical data are treated completely a-historically, with the constant assumption that the narrated events are immediately part of the present, e.g., Abraham set out from Ur of the Chaldeans because he was called *by the God of Ekuphakameni*, the present religious centre of the Nazarites (2:2a). In the same way the Shembe sanctuary is equated to "the heavenly Jerusalem" of Hebrews 12:18ff.

The absence of a "Western" salvational-historical approach in the sermons, coincides with the time concept of oral culture. Time expressions indicate the value which the speaker and the audience attach to what is being said. This is also the reason why oral cultures sometimes express the future in the past tense. Temporal expressions have the pragmatic function of compelling the audience to decisive action in the immediate future. The present and not the past, is the locus where God acts decisively. By their proclamation the salvational events are experienced as present events controlling the present and the future.[7] This oral conception of time explains why Ekuphakameni and the heavenly Jerusalem can be equated. It is interesting to note that Vilakazi criticises Oosthuizen severely for remarking that "Heaven has come down on Zululand" when he evaluates the meaning of the Shembe sanctuary. Vilakazi explains that Ekuphakameni "is a place-name and has little to do with whether the place is on a mountain or on a hill" and that it is "symbolic to the Shembe followers like the City of Jerusalem

---

[7] Cf Ricoeur for his theory of distanciation (1976:43ff).

in Revelation" (1986:111). In the light of oral hermeneutics, neither Oosthuizen's nor Vilakazi's view can be accepted: Ekuphakameni is definitely more that a mere place-name or symbol as Vilakazi intimates; but it would also be wrong to suggest that it completely replaces heaven.

The oral conception of time also explains the way in which tradition is treated, being constantly reappropriated in service of the present.[8]

4. Soteriological significance/authority of the evangelist

It is noteworthy that absolute adherence to Shembe is required of the disciples. It seems the case that *extra Shembe, nulla salus*.

In 1:4 we read that one should stick to Shembe like a mother who refused to desert her dead child. Disciples who persevere in spite of poverty will receive kingship from Shembe (4:1b). The example of Sulile is given, who was martyred for Shembe (2:6-7). The story of Ndlangamandla, a Nazarene Jona, is related in 4:2-3. Deserters are warned that God might reject them (8:1). Shembe requires such absolute obedience that he can even demand that someone parts with a woman (8:1b-2).

Oral theory assists us in understanding this concept of authority that to modern minds seems to border on paternalism. In oral culture knowledge is difficult to attain and therefore precious. Oral cultures are therefore conservative by nature. The wise men (like Shembe and his successors), who are in a position to preserve knowledge, are held in high esteem.[9] Kelber explains with reference to Paul that also his gospel "... echoes the voiceprints of an oral authority" (1983:144,147).

---

[8] It was, e.g., found by Koch that 56% of all citations in Paul have been modified to suit the new context, cf Stanley 1990:78.

[9] Cf Pobee 1989:90 mentions that it was because Paul was bearer of the Jesus tradition that he was put on a level with the other apostles.

The person of the evangelist also has a soteriological significance. There are many indications of this in the text. Reference is made to the perfect example of Shembe and his total commitment to his mission. In 13:3b Shembe leaves his wives to honour his calling. In 1:5 the point is laboured how difficult it was for Shembe to leave his son and wives. In 2:2 the people thought Shembe was mad because he obeyed his calling, but he was led by the Spirit.

The Shembe evangelist, Londa Shembe, preaches to whites although he feared that his action might have caused "the flow of blood." His magnanimity and forgiveness is demonstrated when he bears no grudge against a white lawyer who robbed him of R30 000 (7:1-2).

5. Contextuality/importance of the context

The soteriological significance of the person of the evangelist becomes even more clear when considering the christological language in which the deeds of Shembe are narrated. The descriptions are completely contextualised within the socio-political and cultural world of Natal-KwaZulu. To a western observer it may seem that Shembe is Christ himself, clothed in local African attire.

Among Shembe's many miracles is the healing of an Indian who refused to go to the doctors, witchdoctors and herbalists (2:5). After healing a white child, Shembe refused the money offered to him (10:2). On another occasion Shembe resurrected a white girl after the white *boer* had bound him with ropes and threatened him (7:2b).

The account of Shembe's divine birth is not without resemblance to that of Jesus and John the Baptist (12:3). A childhood narrative tells how God hid him from a *boer* when stealing peaches (12:4-5).

Like the great figures in the Bible, Shembe also had a calling vision. He dreamt that he was flying in the air and saw three men from the north wearing dresses that were shining more than the sun. When Shembe desired to have such a dress, he was told that one can only get such a dress if one behaved well. "Shembe was disappointed, he cried and preferred dying to going back to his dirty body" (12:6-7). This report differs somewhat from Dube's as given by Vilakazi (1986:24).

Similar to Jesus he was also tempted by opponents: a girl is sent to seduce him, but she breaks down and confesses when he compels her to swear on the Bible that God had sent her (15:3).

When he was baptising converts the scene is reminiscent of that of John the Baptist baptizing people. The story runs as follows (15:1):

> White men from Port Shepstone in the company of ministers came and asked him, "Who are you?" He had been baptizing at sea that day. The sea was very rough but he went down with his followers. Some people said he can't do it. He will be risking his own life. There were people there at the beach. Shembe then went into the sea. He got right inside the sea. When he was there, he beckoned to his followers to come. When they came he showed the wave to them, the wave covered the person and the process was repeated until they were all finished.

When Shembe had finished, these whites and these ministers came to ask him, "Of all the prophets, which one are you?" He replied, "I am not Jesus, I am not Jeremiah, nor any other, but I am Shembe."

> The newspapers were already having headlines saying, "Shembe has done a mysterious thing. He baptized people in the sea!" The whites asked him, "Where do you come from because we've been paging the Bible and we've never found Shembe there?" Then Shembe said, "My children, I cannot tell you that. All that I can tell you is that when Moses saw the burning bush, I was there."

Shembe further acquired supernatural dimensions: A dream revealed him to be in heaven (9:6); he is described as divine mediator (12:2). In 14:2b-3 mention is made of his pre-existence. Like Jesus said about himself, people should eat and drink Shembe; "no food is more delicious than Shembe" (7:1). On the day of judgment snakes, stones and little children will testify to his works (15:2).

Oral theory also has an explanation for the contextual nature of the spoken text. In order to retain the attention of the audience, the speaker has continuously to adapt his material to the life situation of the audience. Due to the direct relationship between speaker and audience, oral texts are "speech acts" in the true sense of the word.[10] Their illocutionary and perlocutionary functions are obvious, which is usually not the case with written texts.[11]

Oral communication invariably necessitates a direct contextual involvement of the speaker. It requires that every narrative must be introduced uniquely into a unique situation (Ong 1982a:41). Pre-literate people think situationally rather than categorically (Ong 1982a:52). "The oral mind is uninterested in definitions" (Ong 1982a:47). The meaning of every word is thus controlled by the real-life situations in which the word is used here and now (Ong 1982a:47). A biblical

---

[10] Patte (1983:17-8) presupposes that "any idea or argument ultimately aims at causing us to do something, which might only be a very remote goal . . .." With Paul these goals seem most of the time quite obvious (e.g. 2 Corinthians 8:9 where economic christological metaphors are used in order to encourage financial contributions to the poor). In other cases it may, however, be much more difficult to determine the exact speech act: "If one looks even cursorily at a transcribed record of a conversation, it becomes immediately clear that we do not know how to assign speech acts in a non-arbitrary way" (Levinson 1980:20)

[11] Lyons 1977, 1981:183-8; Ricoeur 1976:27.

example of this is that Paul does not link the "logos" with its content, but with its effect on the hearer - e.g., the "word of reconciliation" in 2 Corinthians 5:19 refers primarily to the reconciliation effected by the gospel.[12]

From the above it is clear that the identification of Shembe and Christ is almost inevitable. It is just as inevitable that a hermeneutic process arises in which the real events in the life of Shembe are constantly communicated in terms reminiscent of Jesus. The focus is never on Jesus or God, but on Shembe as exponent of divinity.

6. Agonistic tone

Ong explains that oral texts are agonistically toned - "orality situates knowledge within the context of struggle" (1982:43-4). Usually proverbs and riddles are used to encourage intellectual combat, while both flyting (reciprocal name-calling) and praise take on dimensions which may seem exaggerated to people from a high-literacy culture (1982:45). The agonistic tone of the sermons is confined to references to Shembe's enemies and the opposition he faced. A short list of some of the references to enmity is as follows:

* Shembe dreams of a spear that would stab him (1:6; 14:1b).
* God scatters the enemies by thunder (3:2b-3).
* The ancestral enemies inside the camp (3:4).
* Beware of demons (4:1a).
* Prayer for deliverance from the enemies (7:3).
* Shembe overcomes violence by a handshake (9:5a,b).
* Prayer that there may not be "cats and dogs" in Ekuphakameni (10:1).
* Many did not love Shembe; in this he resembled Jesus (14:1c-2).
* Hatred among the believers at Mikayideni (6:2a).

---

[12] Kelber 1983:145

Although these topics undoubtedly refer to real threats, the way in which they are presented fits the oral frame of mind. Within the sermons they serve to dramatize the situation in order to maintain contact between speaker and audience.

7. Conceptions of solidarity and individuality

An oral culture deals differently with concepts such as solidarity and individuality than modern cultures. Expressions in an oral culture are empathetic and participatory rather than objectively distanced (Ong 1982a:45, Kelber 1983:150). So, e.g., is Homer's objectivity enforced by formulaic expressions - "the individual's reaction is not expressed as simply individual or 'subjective' but rather as encased in the communal reaction, the communal 'soul'" (Ong 1982a:46 with reference to Havelock). Plato excludes poets from his Republic, because the studying of poems was essentially learning to react with 'soul', i.e. to feel oneself identified with Achilles or Odysseus (Havelock 1963:197-233), something he tried to avoid. In the performance of the West African Mwindo Epic, the narrator slips into the first person when describing the actions of the hero. So bound together are narrator, audience, and character that in one instance the narrator assimilated into his oral performance even the transcribers who were de-oralizing it into text (Ong 1982a:46).

It follows from the above that in an oral culture, the presence of Christ or the Spirit in the church and in the individual believer would be much more comprehensible than in a highly literate culture.[13] Within this frame of mind the presence of the Spirit and faith in Christ and in the future vindication of the believer can only be seen as complementary.

[13] Cf Kelber 1983:145

From the above it is clear why the Nazarites are believed to be in a relationship to Shembe as Paul believed the Christians to be "in Christ." Solidarity with Shembe is possible because he is experienced as a supra-personal spirit communicating with the audience through the preacher. "Shembe who we are following is the love we have in our hearts" (1:5). "Shembe our lord is in heaven where our forefathers are . . . . Shembe says, "Because God loves you, He says I must tell you never to leave this place again" (5:2).

In the sermons the prestige of the group is constantly boosted to enhance group solidarity. The preacher in one place mentions that the meeting at Ekuphakameni surprises those who deserted them (3:1). He reminds them of the wonder of their being together at all (3:4b). They are reminded of their divine election in the story of the Shembe Noah, Ndlangamandla (5:1). Ekuphakameni is said to be heaven itself (5:2a,b). The Nazarenes came from heaven, but it was Shembe that built the house (6:2c). The nobility of Zulu customs is extolled (5:2b). The Nazarenes are gathered at Ekuphakameni because they got permission from the king of Ekuphakameni to be present (12:1). Jesus is with them (3:5). Much emphasis is placed on the uniqueness of the Shembe community: the spirit at Ekuphakameni is unique (1:4/b); there is no other church that can offer a black person such an opportunity to come to God (6:2b).

Through expressions as the above, Shembe himself and his followers, are portrayed as a spiritual unity, a corporate personality, completely in correspondence with the oral frame of mind.

It is interesting to note that some vital aspects of individuality, e.g., the problem of personal guilt, are never addressed in the sermons. Neither is there any indication of the complexity of right and wrong within the individual. Good and evil are sharply distinguished and mainly defined in terms of group adherence.

CONCLUSION

We have found that the Shembe sermons reflect the distinct features of religious communication in an oral culture. In order to understand them, one has to note the oral hermeneutics involved in their production. This is a hermeneutical process underlying much of the biblical text itself. The sermons' non-linear structure, open-endedness, a-historical time conception, the type of authority reflected and their agonistic tone all seem to be unavoidable within such an oral religious culture. Of special interest is the oral conflation of Shembe's deeds and those of Jesus Christ. It has been a major point of criticism that Shembe eventually replaced Jesus Christ (cf Oosthuizen as reported by Vilakazi 1986:89ff). However, Vilakazi's defence of Shembe, which rests mainly on the accusation that Oosthuizen mistranslated the Shembe hymns, does not satisfy and only confuses the issue. What Vilakazi may have felt, was that Oosthuizen did not understand the oral hermeneutics of Shembe christology. From an oral point of view it seems that the portrayal of Isaiah Shembe in christological garb is a natural result necessitated by the medium of communication.

It therefore seems important that, before one can evaluate Shembe preaching in terms of Western dogmatic concepts, one has to understand its oral context. To apply a distinctly "Western" critique, as e.g. a salvational-historical approach, would be unreasonable. The latter is a product of a highly literate culture.

The Shembe sermons do not follow the logic of modern "common sense".[14] Theology in the modern sense of the word is a by-product of literate

---

[14] Cf Bassler 1989:414 "Paul was not a systematic theologian"; Vos 1989:34 refers affirmatively to Raeisaenen's self-critical question whether it would not be anachronistic to analyze a Jew from antiquity from the view point of modern "common sense".

culture and only became possible since the middle ages.[15] Each oral application of Scripture is a "speech act" in the true sense of the word (though their inscripturation obscures their illocutionary force to some extent).[16] In the oral medium there is no other possible way in which one can theologize.[17]

An interesting question is whether the oral matrix is indispensable to the gospel as such. In other words: do people in a high literacy culture need to develop a post-literate orality (Boomershine 1987:152) or to return to a pre-literate understanding of the text? (Lindbeck in Ford 1989:267). Or does the gospel need to be de-oralized so that people in a modern paradigm may understand it better?[18] Ricoeur's positive assessment of the written word and the "distanciation" it entails, points in the latter direction.[19]

---

[15] Furnish 1989:246-7 defines theology as "critical reflection on the meaning and truth of the gospel." He also remarks that it was only among the twelfth century scholastics that theology came to be regarded as a comprehensive, systematic presentation of the gospel. Cf also the criticism against Raeisaenen as cited by Vos 1989:34.

[16] Paul's theology is now widely construed as an activity, cf Bassler 1989:418-9.

[17] Not even the theology of Philo and the rabbinic schools, though comprising of a literary activity, can be seen as "theology" in the modern sense of the word.

[18] Perhaps the orality of the gospel is a reason why in modern societies illiterates, children and women, have a closer affinity for religion than the rest of society. These groups are naturally closer to oral culture.

[19] 1976:37ff "A plea for writing."

BIBLIOGRAPHY

Bassler, Jouette M 1989. "Paul's Theology: Whence and whither? A synthesis (of sorts) of the theology of Philemon, 1 Thessalonians, Philippians, Galatian, and 1 Corinthians," *SBL Seminar Papers*. Atlanta GA: Scholars Press, 412-423.

Boomershine, Thomas E 1987. "Biblical megatrends: towards a paradigm for the interpretation of the Bible in electronic media," *SBL Seminar Papers*, 144-157.

Dewey, Joanna 1989. Oral methods of structuring narrative in Mark. *Interpretation* 43, 32-44.

Ford, David F (ed) 1989. *The Modern Theologians: An Introduction to Christian Theology in the Twentieth Century*. Two vols. Oxford: Blackwell.

Furnish, V P 1989. "Theology in 1 Corinthians: Initial Soundings." *SBL Seminar Papers*. Atlanta GA: Scholars Press, 246-264.

Havelock, Eric A 1963. *Preface to Plato*. Cambridge: Belknap of Harvard University.

------- 1984. "Oral composition in the 'Oedipus Tyrannus' of Sophocles," *NLH* 16, 175-97.

Kelber, Werner H 1983. *The oral and the written Gospel, the hermeneutics of speaking and writing in the synoptic tradition, Mark Paul and Q*. Philadelphia: Fortress.

Kiernan, J P 1990. *The production and management of therapeutic power in Zionist Churches within a Zulu city*, Studies in African health and medicine, vol 4. Edwin Mellen Press: Lewiston.

Lyons, John 1977. *Semantics 2*. Cambridge: Cambridge University Press.

Ong, Walter J 1967 b. *The presence of the Word*. New Haven: Yale University Press.

------- 1977. *Interfaces of the word: studies in the evolution of consciousness and culture.* Ithaca: Cornell University.

------- 1982a. *Orality and Literacy, the technologizing of the word.* London: Methuen.

------- 1982b. "The oral and the written word: a reappraisal" in *The literate revolution in Greece and its cultural consequences.* Princeton: Princeton. 3-38.

Patte, Daniel 1983. *Paul's Faith and the Power of the Gospel, a structural introduction to the Pauline letters.* Philadelphia: Fortress Press.

Pobee, John S 1989. "Oral theology and Christian oral tradition: challenge to our traditional archival concept" in *Mission Studies*, Journal of the IAMS, 6(1), 87-98.

Raeisaenen, H 1986. *Paul and the Law.* Philadelphia: Fortress.

Ricoeur, P 1976. *Interpretation theory: discourse and the surplus of meaning.* Fort Worth, Texas: Texas Christian University Press.

Stanley, Christopher D 1990. "Paul and Homer: Greco-Roman citation practice in the first century CE," *Novum Testamentum* 32 (1), 48-78.

Vansina, Jan 1965. *Oral tradition, a study in historical methodology,* translated form French by H M Wright. London: Routledge & Kegan Paul.

Vilakazi, Absolom with Bongani Mthethwa & Mthembeni Mpanza 1986. *Shembe, the Revitalization of African Society.* Johannesburg: Skotaville Publishers.

Vos, J S 1989. "Nieuw licht op de apostel Paulus, tendenties in het huidige onderzoek," *GTT* 89(1), 23-40.

CHAPTER 18

FROM VENERATION TO DEIFICATION OF ISAIAH SHEMBE: REFLECTIONS ON AN ORAL HISTORY

by

M.C. Kitshoff

A document given to Professor G C Oosthuizen by Londa Shembe a few years ago contains the precipitation of an oral history of how a Sotho family in the land of Moshoeshoe was led to Isaiah Shembe in Ekuphakameni. Apart from this amazing and even moving account of their trek to Durban, the narrative also sheds interesting light on the extent and intensity of the veneration with which Shembe, founder of the Church of the Nazarites was treated. Our reflections will mostly be directed at the latter.

SHEMBE AND THE PEOPLE FROM LESOTHO

In 1918, the year of the influenza epidemic, Nyathi Petros Hadebe and his family, from the district of Mphavane, had an extraordinary experience. It was revealed to them, through visions and dreams, that Jesus had returned to the world and was living among black people in the form of a black man. Nyathi and his house were advised to go and see him.

A herbalist, Moses, who visited the family, was told about this revelation. He assured the family that there was a prophet in Durban, but, in his opinion, that man was not Jesus.

Hadebe was determined to go and find this man. Two of his three wives and children were willing to accompany him. Among the children was the eyewitness who recalled the story of their search for the prophet.

The journey to Durban was one of hardship and grief. One of Nyathi's wives and three of the children died of influenza. His other wife, Melika, daughter of Chief Smith Ngwenya, lost heart and wanted to go back. But Nyathi persisted. In their state of destitution they were even forced to steal food from the Boers. A dream by Melika that she was being baptised by a man who resembled Jesus encouraged them and gave evidence that Nkulunkulu (God) was with them on their way.

At last they came to Durban, but their situation became one of complete deprivation and hopelessness. In his state of suffering Nyathi even went so far as to sell his daughter to a white man. A sympathetic stranger gave him an address and advised him to go there. The details were: Rev J L Dube, Ohlange, Phoenix, Natal.

When he told the people at Dube's place about the dreams and visions, they wondered whether those revelations were not referring to Shembe and his village, Ekuphakameni. They were, however, not very encouraging, but called the followers of Shembe evil people who alleged that Shembe was Jesus, and who spoke of him as though he was Nkulunkulu.

But nobody and nothing could dissuade Nyathi Petros Hadebe. He was determined to meet Shembe. And his determination was rewarded. During June 1919 he met the "prophet of Jehovah" who spoke to him in Sotho. His impression of Shembe was: "This man is awe-inspiring to look at; he has a great presence. And yet he does not resemble the people of this earth".

Hadebe and his wife and children soon became believers and were baptised by Shembe, exactly as Mother Melika had dreamt. Shembe took them all in at Ekuphakameni and later settled them at a place called Enhlanzini.

This then is briefly the story of a Sotho family who were the first people from the other nations who came to Ekuphakameni - and found a place to feel at home.

When one reads the narrative under review, one becomes aware of the profound reverence towards Isaiah Shembe which culminates in adoration, veneration and even deification. To these aspects more attention is to be given now. In the discussion some references will be made to the study of Absolom Vilakazi, with Bongani Mthethwa and Mthembeni Mpanza, *Shembe - The Revitalization of African Society*. Johannesburg, Skotaville, 1986, which also dwells on what is called "the sacralization of Shembe".

SHEMBE AND BIBLICAL PARALLELS

The reverence towards Shembe as reflected in the narrative is often couched in terms reminding the reader of biblical events or biblical ideas. Throughout the narrative these biblical parallels present themselves. Sometimes they are directly pointed out by the original narrator or the editor of the narrative; sometimes parallels can be detected by the reader of the narrative who possesses some understanding of the Bible. In the latter case the narrator himself might even not have been conscious of such a parallel between his narrative and the Bible.

What might be an unintentional parallel is found at the very beginning of the narrative. It is told how on a certain night, at midnight, the sound of beating of drums and of singing was heard, and all the music came from the sky. Then a round object appeared in the sky and glowed beautifully with a pale light. When the music ceased and the luminous object disappeared, amazement and fear, but also joy, filled the hearts of the beholders.

The biblical parallel seems obvious. When, according to the Gospel of Luke, Chapter 2, Jesus was born, there was singing in the air produced by heavenly messengers, while the light of the glory of the Lord shone all around the shepherds. They were terribly afraid but when they heard the good news about the Messiah who was born, great joy came over them.

Whatever the case may be, whether a narrative deliberately constructed to parallel the events on the plains of Bethlehem, or a mere narrative about an experience on the highlands of Lesotho, this narrative from the outset placed the main actor, Isaiah Shembe, in the niche of someone very special. Against this background where the natural and the supernatural meet and blend, evidence is given of a person very dear and very near to God. Although the narration of the wonderful signs in the sky and the deep emotions in the hearts cannot be construed as a prelude to an eulogy, a reader of the narrative can in retrospect interpret this story as signifying divine action and divine interest in, if not divine approval of the person and work of Isaiah Shembe. That is why the recording of this event is preceded by this statement, "The manner of the coming of the Word of Jehovah in the land of Lesotho was in this way".

The expression "coming of the Word of Jehovah", cannot but turn the reader's mind to the claims found in many of the opening remarks of Old Testament prophecies, namely that "the Word of God came" to the speakers. By using this prophetic formula they claimed divine guidance, divine authority, divine legitimation and, at times, even divine compulsion. And by all these they asserted that they were reliable mouthpieces of God.

The experience of the drum-beating and singing in the sky, the bright light in the night and the fear and joy in the heart, came to be explained by a vision recorded in the narrative. As a result of this visionary communication, the parallel sensed or assumed by the reader, namely a parallel with the events in the fields of Bethlehem, now becomes more explicit. A voice spoke to Melika, the wife of Nyathi Hadebe, and told her that the sound of drums and the singing they had heard were connected with Jesus who had returned to live "among Black people ... in the form of Black people". This announcement very much sounds like an Africanized version of John 1:12,14 where we read that the Word became a human being who came to live among his own people. The voice urging Hadebe and his house to go and see this man in the east also reminds the reader of the shepherds of Bethlehem who on receiving the message of the newly born Saviour, could not do otherwise than going to "see this thing that has happened, which the Lord has made known to us" (Luke 2:16).

The revelation of this "thing" which came to them verbally and visually was experienced by the shepherds as coming from the Lord. With Hadebe it was not otherwise. He, as reported by the narrator, understood the visions and dreams as the "coming of the word of Jehovah", and as Unkulunkulu speaking to him.

A further direct parallel between the oral history under review and the biblical narrative is contained in another vision of Melika. She saw that she was being baptised by a man, who later turned out to be Isaiah Shembe. She recalled: "When I looked at this man he resembled Jesus, for he had wounds on his feet and on his hands. I could not see whether he had any wounds on his side for his side was concealed by the robe that he worn. But his hair were pitch black".

This is a very revealing parallel, revealing not only knowledge of the biblical crucified Jesus, but also revealing the extent of the veneration paid to Shembe, veneration which extends to jesufication and even perhaps deification. To this matter will be returned later.

A remark by Hadebe's wife brings a further parallel to light. When they at last reached Ekuphakameni, the village of Isaiah Shembe, Hadebe was confused because the village was so small. But his wife reminded him about the village of Bethlehem: though insignificant, it was there that Jesus was born. The obvious parallel is that both Shembe and Jesus were great persons associated with small villages, but there is certainly something more to it. Here again we have veneration indirectly expressed, but the impression is gained by the reader that this veneration borders on the identification of Shembe and Jesus.

SHEMBE AND HIS DESIGNATIONS

That Isaiah Shembe was revered, even adored, as a very special person cannot be doubted. Vilakazi in this respect speaks of "the sacralization of Shembe" by his followers (1986:36-41).

This veneration, in various degrees, is very often reflected in the entitling of Shembe and the ways in which he was addressed. The narrative under review gives ample evidence of this.

The following designations are used:

the Nkosi of heaven
the Nkosi of the AmaNazaretha
the Nkosi of AmaKhosi
the great prophet
the messenger of Nkulunkulu
the son of Nkulunkulu
the man of Nkulunkulu
the lord of mercy
the man of heaven
the great hero of heaven
the wonderful man
the wonder of wonders
the extraordinary man

All of these titles and descriptions reveal veneration, though in various degrees. We shall look at the following designations in more detail:

(a) The great prophet

This designation, used thrice in the narrative, expresses veneration commonly held by the followers of Shembe. Vilakazi sees the veneration verbalized in the title "prophet" as part of the process of the sacralization of Shembe. He follows Max Weber and contends that the vestment of authority based on personal revelations and charisma is the hallmark of a prophet. A prophet should, according to Vilakazi, also practise healing and counselling without charge. That Shembe did.

Biblically speaking the essence of the task of the prophet is to speak God's words, to make God's will known. Our narrative gives little evidence that Shembe did that. However, Vilakazi has no doubt in his mind that "Shembe was a prophet in the nature of the biblical prophets. He called on God and was always enlightened by God according to his own sayings and the evidence of his followers" (1986:104).

Conceded that there are parallels between Shembe and prophets in the Bible, the real issue revolves round the title of "Great Prophet" ascribed to Shembe. Should one look for a biblical parallel it would only be found in Luke 7:16 where this title expressed praise and awe towards Jesus Christ for raising the child of the widow of Nain.

While the title, great prophet, designated to Shembe gives evidence of respect, honour, admiration and awe, in short reverence, there is no sound reason to claim that this title of Shembe, which is the same as one of the titles accorded to Jesus Christ, should be seen as an example of jesufication. However, when one reflects on the content and tenor of the narrative as a whole such a possibility is not to be ruled out completely.

(b) The son of Nkulunkulu

In one of the visions of Melika mention is made of "holy men" who dwell in the east in a village which was built by the "son of Nkulunkulu". This statement is followed by the explanation that Jesus had returned to the world and was living in that village.

The term "son of God" can certainly be understood in a general sense of being created by and cared for by God, or in a biblical sense of being a child of God by faith and obedience (John 1:12, Luke 6:35).

However, it does not seem as if only these two meanings apply to Shembe's title. The doubts are strengthened by the context in which the title is used. By identifying the "son of Unkulunkulu" with Jesus, this term can certainly be construed by the reader as an example of jesufication.

Furthermore, Jesus Christ is also known as the Son of God. Theologically this can be interpreted that the Son is co-essential and co-existent with the Father and, therefore, co-divine. When Shembe is seen as an identification or manifestation of Jesus, and when Jesus is confessed as being divine, the title son of Unkulunkulu can certainly be perceived as conveying evidence of deification as well as veneration.

(c) The Nkosi of heaven and related titles

As Vilakazi reminds us "iNkosi" means "ruler", "leader", "chief" or "king". But it can also mean "God". The meaning must be derived from the socio-cultural context. In this connection Vilakazi accuses Oosthuizen of following a "dictionary approach" and operating from "theological prejudices" when Oosthuizen argues that Isaiah Shembe regarded himself as God (1986:63-64, 96-97).

While rejecting the idea that Shembe saw himself as God, Vilakazi explains why Shembe is referred to as iNkosi: "Shembe is not merely a leader in the church, he is the ruler, the person who lays down the law, just as a father does in the home, a king in national affairs" (1986:63). Vilakazi further argues that Shembe's position as chief was enhanced by the fact that he was also prophet and healer. These activities raised his position to that "equal to a king, for a king was the biggest medicine man in the land (1986:64).

One need not disagree with Vilakazi's socio-cultural explanation, but a problem arises when Shembe is called "iNkosi of heaven". Certainly this designation can be understood that Shembe the chief and ruler was guided "from above" or that he received his power and authority from God in heaven. For the reader of this narrative, however, who is reminded by the Bible that heaven is the abode of God (Matthew 6:9) and that Jesus was the person who came from heaven (John 6:33-51), and who is reminded by the narrative itself that Shembe was nobody else than Jesus who had returned, this title assumes proportions which cannot be adequately explained in socio-cultural terms. The title "Nkosi of heaven" and even the more neutral "man of heaven" or "great hero of heaven" expands the context to include the religious, but also communicates the notion of an authority which is not vested in a human being. These titles, therefore, not only verbalize an intense veneration but are expressive of, or are opening doors for jesufication and deification.

A related title of Shembe as appeared in the narrative is "iNkosi of the AmaNazaretha." Undoubtedly the title confirms, as Vilakazi asserts, that Isaiah Shembe, a ruler of his group, "has the final word on all matters, and he pronounces final judgments and dispensations. All contentious matters are referred to him for his final ruling. He thus governs the church as a temporal king governs, and dispenses justice among his subjects" (1986:64). To all this one can agree, but when this title is read in conjunction with such expressions as

"Nkozi of heaven" or "man from heaven", one is certainly justified in not summarily dispelling the impression that even this title contains elements of deification.

This brings us to the last title in this category: "Nkosi of Amakhosi" (as it was translated). Vilakazi assures us that in a religious context no Zulu would ever understand the address, "Nkosi Yamakhosi" (king of kings), as referring to Isaiah Shembe (1986:97). In the narrative under discussion this designation is not used in a religious context but is referring to Isaiah Shembe. Read together with designations discussed under this paragraph one becomes convinced that "king of kings" as a title of Shembe is more than a reference to or praise and respect for an leader, king or ruler who surpasses all others in high offices.

Furthermore, when the Christian reader of the narrative confesses on biblical grounds that God is the "King of kings and Lord of lords" (1 Timothy 6:14) and that his Son, Jesus Christ is also "King of kings and Lord of lords" (Revelation 19:16) such a reader cannot but conclude that the title "King of kings" when directed at Shembe is expressing high-keyed veneration, but is also, at times, a vehicle of jesufication and deification.

SHEMBE AS THE THREE-IN-ONE AND PRE-EXISTENT ONE

A son of Nyathi Petros Hadebe, who like his father also became a follower of Shembe and who, in all probability, became the originator of the narrative under discussion, also received a vision. This son, Meshack Hadebe, saw Shembe as having three bodies. The first body was in the middle of the earth but up in the sky, the second one was in the north at the edge of the sky, and the third one was in the east also "on the very edge of the firmament of the sky". The whole being was "clothed with the sun" and enwrapped in glory beyond description.

Then the "voice of voices" came to Meshack and said", "Do you now realize who Shembe is? Listen. At a time before your father Nyathi was born, Shembe was already existing; he was a man of mature age even as you now see him. He was as you see him in these three beings who are him. I say to you that before your father was born Shembe was already existing and that he was even then as you see him now".

When reading about this vision in which Shembe appeared in three bodies, this three-in-oneness cannot but call to mind associations with the Christian doctrine of the divine trinity. Although this teaching can in no way be adequately and logically explained, the fundamental issue is: God is one but he revealed himself in Scripture as three persons. But whereas Shembe as one person is revealed in three bodies, God who is one, and who is Spirit and who has no body, is Scripturally revealed in three persons.

This vision of Shembe as a three-in-one being, certainly does give the reader who agrees with the teaching of the divine trinity reason to believe that such a picture of Shembe does reflect, to a certain extent, a deified Shembe.

This belief that Shembe is deified is given confirmation when the "voice of voices" alluded to the pre-existence of Shembe. Paraphrasing the content of the vision and voice, one can say that the vision claimed that Shembe was not one-body-bound nor space-bound, while the voice asserted that Shembe was not time-bound nor change-bound. From this one may conclude that the combined contention of vision and voice was that Shembe, if not like God Eternal, at least revealed some essential divine attributes.

At this stage in the written version of the story scriptural references are given, most probably inserted by an editor. They are John 8:56-58, Proverbs 8:23,25,27 and Hebrew 7:2-3. The reference in John pertains to the pre-existence of Jesus Christ, in Hebrews to Melchizedek who was an eternal priest, like Jesus Christ, and in Proverbs to the pre-existence of Wisdom, which in New Testament light is often seen as pointing to Jesus Christ.

It could be that the editor had no other intention than to indicate interesting parallels between Shembe and Jesus. The sensitive Bible reader, however, will in all probability conclude that the combined weight of vision, voice and the selected Bible references tips the balance heavily in favour of the view that the person of Isaiah Shembe in this narrative is made the object of jesufication and deification.

SHEMBE AS CO-CREATOR

In the narrative under consideration we heard about an amazing self-revelation of Shembe during June 1924. Shembe instructed a group of people to clear a site for a temple and to dig a hole for a baptismal pool. He told them that they would have to dig until they found six round stones. Then he uttered these astonishing words: "I put these stones there on the day my father and I created the earth". While the workers were still reeling under the impact of this disclosure, Shembe told them that below those six stones they would find a rock, and said, "this was placed in that place by myself at the time when my father and I created this earth. I personally placed this rock in this place". This claim, according to the narrator, was repeated a few days later.

The narrator related how he in unbelief pondered this mystery too profound to fathom, and how the piercing eyes of Shembe rebuked him for his unbelief. On 10 June 1924 the first round stone was unearthed after they had to dig through soil as hard as a rock. Then Shembe turned to Meshack Hadebe and said, "Do you now believe?" Meschack affirmed his faith.

At this point the narrative, through its narrator or editor, refers to John 3:2, where Nicodemus acknowledged that Jesus came from God otherwise he (Jesus) would not be able to do the signs he was doing. The point which the narrator or editor wanted to make was obviously that Shembe came from God.

And if he came from God, it could reasonably well be accepted that he was with the Father on the day of the creation.

But there is more to it. John 3:2 refers to Jesus who, according to Nicodemus, came from God. And Jesus himself confirmed this at another occasion (John 6). Moreover Jesus Christ as Son of God is also seen in the Bible as one who co-operated in the process of creation (John 1:3, Hebrew 1:2-3, Colossians 1:15-17).

This parallel between Jesus Christ and Isaiah Shembe where both are seen as active agents at the creation, might not have been visualized or intended by the narrator or editor. Neither is it transparent that by referring to the Bible passage mentioned, the intention was to identify Shembe with Jesus. What is beyond doubt is that both the claims by Shembe as recorded by the eyewitness and the Bible passage quoted in the narrative make a strong appeal to recognize and venerate Shembe as co-exister and co-creator with God the Father. The sensitive Bible believer will also hear this appeal followed by the sounds of enthusiastic veneration. But while having his ear to the Bible, he will most probably in the veneration hear the tone of jesufication and deification.

CONCLUSION

Shembe was a leader of exceptional qualities - a genius. The exuberant reverence paid to him is understandable. But in the stream of reverence Shembe's followers often found themselves in the rapids of jesufication and deification.

In our reflection on an oral history we have ascertained that instances of jesufication and deification to a greater or lesser extent presented themselves in the narrative. These examples could perhaps be subsumed under Vilakazi's term: sacralization.

Vilakazi rejected the notion that Isaiah Shembe projected himself as God or Saviour. He admitted that Shembe was thought to have a divine nature, but argued that that idea was derived from life experiences and testimonies of his followers (1986:92,115). Regarding Shembe and Jesus Christ, Vilakazi says: "Christian complaints that Shembe I usurped Jesus Christ's place are based on a cultural misunderstanding". It seems as if the "misunderstanding" is the failure to understand "the Zulu social system in which the son is not an equal but an extension of the father" (1986:93). But this explanation does not appear to be relevant to the issue of identification with or usurpation of Jesus Christ.

The narrative which we reviewed is a testimony by a follower or followers of Shembe, and as such does not produce conclusive evidence that Shembe ascribed divine attributes to himself or that he usurped the place of Jesus Christ. Even where it is recalled that Shembe claimed pre-existence and co-creatorship with the Father, these claims cannot be verified. The introduction of Shembe speaking in the first person and making divine claims could be a hagiographic technique used by the narrator or editor.

One can conclude with Vilakazi that it would be "unfair to blame Shembe I for what people, who came under his influence, said about him" (op.cit.:115). But perhaps the real point is: Did Shembe reject or resist this sacralization which even during his lifetime reached such proportions as to issue out in jesufication and deification?

Hymn 5 of Shembe's hymn-book, **Izihlabalelo ZamaNazaretha** reads, inter alia;

> Thou who seek Life
> in Jesus' herd
> remember the narrow path.

These words are, no doubt, referring to Matthew 7:13, but will, most probably, also stir up associations with the pronouncement of Jesus Christ in John 14:6, "I am the way". Seen in this light, Vilakazi is justified in saying: "There is no suggestion in this hymn of a desire to denigrate Christ". (op.cit.:94).

But as we have seen, Shembe's followers were confessing more than Shembe himself, and they were contexualizing more than their leader. In our study we have reflected on evidence of the remoulding of Shembe to resemble Jesus. And in this process they have perhaps not wilfully denigrated Jesus, for they gave him a contextualized face - the face of Shembe. But is the new face of Jesus not perhaps a denigrated face?

The Bible declares: Jesus is the way.
The sticker on the rear window of many cars in
Kwazulu-Natal replies: Shembe is the way.

Substitution, parallelization, identification, jesufication or usurpation? Which?

There is however, another way of understanding the titles and attributes ascribed to Shembe. Perhaps the language used in the document under review is that of praise-singing where symbolism, metaphors and superlatives usually abound. One may, of course, react by pointing out that the names and characteristics with which Shembe is adored were not expressed in the context of praise-singing but as part of an unemotional narrative.

Conceded, but in oral tradition it is easy enough for a great man to grow even greater in stature and power as the story is passed on from mouth to ear. In a religious tradition a person can easily become a saint, and stories about saints usually contain strong elements of hagiolatry, and hagiolatry cannot but have praise-singing as its essence. This seems to be true of many Shembe stories.

Usually the veneration and adoration does not reach higher than hagiolatry. But it is conceivable that the desire for an own divine incarnation - a black Messiah - and the desire to have the fullness of the Spirit manifested in one of their own people, all which coincided with the appearance of charismatic Shembe, could have shifted the boundaries of hagiolatry into the realm of deification and theolatry.

It seems as if the narrative under review in this paper supports this contention.

SCRIPTURE INDEX

NAME INDEX

WORD INDEX

AFRICAN STUDIES

1. Karla Poewe, **The Namibian Herero: A History of Their Psychosocial Disintegration and Survival**
2. Sara Joan Talis (ed. and trans.), **Oral Histories of Three Secondary School Students in Tanzania**
3. Randolph Stakeman, **The Cultural Politics of Religious Change: A Study of the Sanoyea Kpelle in Liberia**
4. Ayyoub-Awaga Bushara Gafour, **My Father the Spirit-Priest: Religion and Social Organization in the Amaa Tribe (Southwestern Sudan)**
5. Rosalind I. J. Hackett (ed.), **New Religious Movements in Nigeria**
6. Irving Hexham, **Texts on Zulu Religion: Traditional Zulu Ideas About God**
7. Alexandre Kimenyi, **Kinyarwanda and Kirundi Names: A Semio-linguistic Analysis of Bantu Onomastics**
8. G. C. Oosthuizen, (*et al*), **Afro-Christian Religion and Healing in Southern Africa**
9. Karla Poewe, **Religion, Kinship, and Economy in Luapula, Zambia**
10. Mario Azevedo (ed.), **Cameroon and Chad in Historical and Contemporary Perspectives**
11. John E. Eberegbulam Njoku, **Traditionalism Versus Modernism at Death: Allegorical Tales of Africa**
12. David Hirschmann, **Changing Attitudes of Black South Africans Toward the United States**
13. Panos Bardis, **South Africa and the Marxist Movement: A Study in Double Standards**
14. John E. Eberegbulam Njoku, **The Igbos of Nigeria: Ancient Rites, Changes and Survival**
15. Aliyu Alhaji Idrees, **Domination and Reaction in Nupeland, Central Nigeria: The Kyadya Revolt, 1857-1905**
16. Kenoye Kelvin Eke, **Nigeria's Foreign Policy Under Two Military Governments, 1966-1979: An Analysis of the Gowan and Muhammed/Obasanjo Regimes**
17. Herbert Ekwe-Ekwe, **The Biafra War: Nigeria and the Aftermath**
18. I. D. Talbott, **Agricultural Innovation in Colonial Africa: Kenya and the Great Depression**

29. G. C. Oosthuizen and Irving Hexham (eds.), **Afro-Christian Religion at the Grassroots in Southern Africa**

20. Bessie House-Midamba, **Class Development and Gender Inequality in Kenya, 1963-1990**
21. David Owusu-Ansah, **Islamic Talismanic Tradition in Nineteenth-Century Asante**

22. Segun Gbadegesin (ed.), **The Politicization of Society During Nigeria's Second Republic, 1979-1983**

23. Tukumbi Lumumba-Kasongo, **Nationalistic Ideologies, Their Policy Implications and the Struggle for Democracy in African Politics**

24. Getachew Metaferia and Maigenet Shifferraw, **The Ethiopian Revolution of 1974 and the Exodus of Ethiopia's Trained Human Resources**

25. George Cotter, **Proverbs and Sayings of the Oromo People of Ethiopia and Kenya with English Translations**

26. J. W. T. Allen, Roland Allen, N. Q. King, Jan Knappert, et al. (eds.), **A Poem Concerning the Death of the Prophet Muhammad /** ***Utendi Wa Kutawafu Nabii***

27. Salah M. Hassan, **Art and Islamic Literacy Among the Hausa of Northern Nigeria**

28. Samuel S. Simbandwe, **A Socio-Religious and Political Analysis of the Judeo-Christian Concept of Prophetism and Modern Bakongo and Zulu African Prophet Movements**

29. Pyare L. Arya, **Structure, Policies and Growth Prospects of Nigeria**

30. Sheikh Batu Daramy, **Constitutional Developments in the Post-Colonial State of Sierra-Leone 1961-1984**

31. Abiodun Goke-Pariola, **The Role of Language in the Struggle for Power and Legitimacy in Africa**

32. Marjorie H. Stewart, **Borgu and its Kingdoms: A Reconstruction of Western Sudanese Polity**

33. Abdalla Uba Adamu, **Reform and Adaptation in Nigerian University Curricula 1960-1992: Living on the Credit Line**

34. Thomas M. Shaw, **The Fulani Matrix of Beauty and Art in the Djolof Region of Senegal**

35. Toyin Falola, A. Ajayi, A. Alao, B. Babawale, **The Military Factor in Nigeria 1966-1985**

36. Samba Diop, **The Oral History and Literature of the Wolof People of Waalo, Northern Senegal: The Master of the Word (Griot) in the Wolof Tradition**

37. Rina Okonkwo, **Protest Movements in Lagos, 1908-1930**

38. Felton Best (editor), **Black Resistance Movements in the United States and Africa, 1800-1993**

39. Jeffrey C. Stone, **A Short History of the Cartography of Africa**

40. H.L. Pretorius, **Historiography and Historical Sources Regarding African Indigenous Churches in South Africa**

41. Michael Okema, **Political Culture of Tanzania**

42. Alven Makapela, **The Problem with Africanity in the Seventh-Day Adventist Church**

43. Patrick E. Idoye, **Theatre and Social Change in Zambia: The Chikwakwa Theatre**

44. MC Kitshoff (editor), **African Independent Churches Today: Kaleidoscope of Afro-Christianity**